CONTENTS

ABBREVIATIONS
v

INTRODUCTION
ix

ART AND WAR
xi

CATALOGUE OF TREATISES
1

ABBREVIATIONS

ADB
Allgemeine deutsche Biographie, (Berlin [1875], 1967).

Architekt und Ingenieur
Architekt und Ingenieur Baumeister in Krieg und Frieden (Wolfenbüttel, 1984).

Bagrow
Leo Bagrow, *History of Cartography,* revised and enlarged by R. A. Skelton, (Chicago, 1966).

Blanchard
Anne Blanchard, *Les Ingénieurs du "Roy" de Louis XIV a Louis XVI* (Montpellier, 1979).

Blomfield
Reginald Blomfield, *Sébastien le Prestre de Vauban, 1633–1707* (London, 1971).

BM
British Museum

BN
Bibliothèque Nationale Paris

BNB
Biographie Nationale de Belgique, publiée par l'Académie Royale (Brussels, 1866)

BRT
Biblioteca Reale Turin

Brunet
J. C. Brunet, *Manuel du libraire et de l'amateur de livres* (Paris, 1860–1865).

BWN
Biographisch Woordenboek der Nederlanden, eds. K. J. R. Van Hardenwijk and C. D. J. Schotel (Amsterdam, 1969)

Cockle
M. J. D. Cockle, *A Bibliography of English Military Books up to 1642 and of Contemporary Foreign Works* (London, 1957).

Dainville
François de Dainville, *L'Education des jésuites, XVIe–XVIIIe siècles* (Paris, 1978).

DBF
Dictionnaire de biographie française, eds. Balteau, Barroux and Prevost (Paris, 1932–1975)

DBI
Dizionario Biografico degli Italiani, Istituto della Enciclopedia Italiana (Rome, 1960–ongoing).

de la Croix
Horst de la Croix, "The Literature on Fortification in Renaissance Italy," *Technology and Culture,* 4 (1963):30–50.

Dezzi-Bardeschi
Massimo Dezzi-Bardeschi, "Su A. C.," *Studi secenteschi,* 4 (1963):45–79.

DNB
The Dictionary of National Biography, eds. Sir Leslie Stephen and Sir Sidney Lee (Oxford, 1959–60).

Duffy
Christopher Duffy, *The Fortress in the Age of Vauban and Frederick the Great, 1660–1789,* Siege Warfare volume 2 (London, 1985).

Early Military Books
Thomas M. Spaulding and Louis Karpinski, *Early Military Books in the University of Michigan Libraries* (Ann Arbor, 1941).

Encyclopedia of Architects
Adolf K. Placzek, ed., *Macmillan Encyclopedia of Architects* (New York, 1982).

EUI
Enciclopedia Universal Ilustrada (Barcelona, n.d.).

Fortifications
Simon Pepper and Nicholas Adams, *Firearms and Fortifications: Military Architecture and Siege Warfare in Sixteenth-century Siena* (Chicago, 1986).

Grodecki
L. Grodecki, "Vauban Urbaniste," *Bulletin de la Societé d'étude du XVIIe Siècle*, 34 (1957):329–52.

Guarnieri
P. E. Guarnieri, *Breve biblioteca di architettura militare* (Milan, 1803).

HAB
Herzog August Bibliothek Wolfenbüttel.

Hale
John R. Hale, "Francesco Tensini et les fortifications 'post-Renaissance' en Venetie," in Martha D. Pollak, ed., "Architecture militaire," *VRBI* 11 (1989):36–55.

HCL
Ruth Mortimer, *Italian 16th Century Books*, Harvard College Library Department of Printing and Graphic Arts (Cambridge, 1974).

Lallemand
Marcel Lallemand and Alfred Boinette, *Jean Errard de Bar-le-Duc: sa vie, ses oeuvres, sa fortification* (Paris, 1884).

Manno
Alessandro Biral and Paolo Morachiello, *Immagini dell'ingegnere tra quattro e settecento; filosofo, soldato, politecnico;* with an annotated bibliography by Antonio Manno (Milan, 1985).

Marini
Luigi Marini, *Architettura militare di Francesco de' Marchi* (Rome, 1810).

Michaud
M. Michaud, *Biographie universelle, ancienne et moderne* (Paris, 1854, and supplements).

Miscellanea 1
Carlo Promis, "Vita di Gerolamo Maggi d'Anghieri, ingegnere militare, poeta, filologo, archeologo, giurisperito del secolo XVI," *Miscellanea di Storia Italiana*, 1 (1860):105–43.

Miscellanea 4
Carlo Promis, "Gl'ingegneri e scrittori militari bolognesi del XV e XVI secolo," *Miscellanea di Storia Italiana*, 4 (1863):579–690.

Miscellanea 6
Carlo Promis, "Gl'ingegneri italiani della Marca d'Ancona che operarono e scrissero dall'anno 1550 all'anno 1650," *Miscellanea di Storia Italiana*, 6 (1865):241–356.

Miscellanea 12
Carlo Promis, "Gl'ingegneri italiani che operarono o scrissero in Piemonte dal 1300 al 1650," *Miscellanea di Storia Italiana*, 12 (1871):411–646.

Miscellanea 14
Carlo Promis, "Biografie di ingegneri italiani dal secolo XIV alla metà del XVIII," *Miscellanea di Storia Italiana*, 14 (1874):1–858.

Parent and Verroust
Michel Parent and Jacques Verroust, *Vauban* (Paris, 1971).

Pastoureau
Mireille Pastoureau, *Les Atlas français XVIe–XVIIe siècles: répertoire bibliographique et étude* (Paris, 1984).

Rocchi
Enrico Rocchi, *Le origini della fortificazione moderna* (Rome, 1894).

Vivenza
Gloria Vivenza, "Giacomo Lanteri da Paratico e

il problema delle fortificazioni nel secolo XVI," *Economia e Storia,* 22 (1975):503–538.

Waetzoldt
Wilhelm Waetzoldt, *Dürer and His Times* (London, 1950).

Wiebenson
Dora Wiebenson, ed., *Architectural Theory and Practice from Alberti to Ledoux* (Chicago, 1982).

INTRODUCTION

IN NOVEMBER 1991, THE TENTH SERIES OF KENNETH NEBENZAHL, JR. LECTURES in the History of Cartography was held at The Newberry Library in Chicago. The theme was "Profiling the City: Six Studies in Urban Cartography." As we were discussing the accompanying exhibit, it became clear that this was an opportunity to compose not only an exhibit-catalog, but also a checklist of the Library's holdings of treatises on military architecture, which had a profound effect on early modern urban planning.

Martha Pollak of the University of Illinois at Chicago agreed to compile such a checklist, which we present here. The work is arranged alphabetically by author, and a full description is given of the Library's copy of each treatise, together with some indication of where other copies may be found. We have included numerous plates, in the hope of giving readers some idea of the nature of each work, and so enticing some of them to undertake research in this generally neglected body of publications.

The book has been designed for us by Robert Williams, of The University of Chicago Press, using photographs made by Kenneth Cain at the Library. We hope that readers will find it both interesting and useful.

David Buisseret
1991

ART AND WAR: RENAISSANCE AND BAROQUE TREATISES ON MILITARY ARCHITECTURE

The Ubiquity of War

"The most beautiful aspect of Architecture is surely that which deals with cities" wrote Pietro Cataneo in the dedication of his treatise to Enea Piccolomini, a descendant of Pope Pius II who had been such a great city builder and lover of architecture. "But, he continued, since cities are now threatened by artillery which the ancients did not possess, I will demonstrate how to build cities differently from theirs so as to defend them from a menace that was previously unknown."[1] Since the association between architecture and the city had been made earlier in published and manuscript treatises by such fifteenth-century writers as Alberti, Filarete and Francesco di Giorgio Martini, the novelty of Cataneo's idea consisted in the further association of city planning with military architecture. But by defining the defense of the city as the domain of the architect in print he merely made a formal claim for a practice already accepted two generations earlier. Thus when in 1482 Leonardo da Vinci offered his services to Ludovico Sforza, duke of Milan, he recommended his own skills in a ten-point letter where he bunched together his acknowledged artistic achievements in painting, sculpture and architecture as the last point, using the first nine to describe enticingly for the duke the war machines and ferocious firearms that he had invented. He commended his own abilities in war, claiming that he could destroy "every 'rocca' or other fortress" while during peace he would be "the equal of any other in architecture and the composition of buildings public and private.[2] When Leonardo, with Francesco di Giorgio Martini and Alberti, claimed military architecture for the Renaissance designer and artist, he bound together two seemingly unreconcilable activities: building and destruction. Furthermore, he associated these activities with that of surveyor and cartographer and he may have been the first artist and military architect to produce an accurately measured and drawn plan of a city—that of Imola—which became a fundamentally

1 *I quattro primi libri* (Venice 1554), p. 1: "... la più bella parte dell'Architettura certamente serà quella, che tratta delle città ... le quali essendo modernamente offese dalle artiglierie, che no havevano gli antichi; non serà presontione la mia, se io mostrerò di edificarle altrimenti, per difenderle da quelle offese, alle quali essi non hanno potuto provedere, per non haverle havute al tempo loro."

2 Ludwig Heydenreich, "The Military Architect," *The Unknown Leonardo*, ed. Ladislao Reti (London, 1974), 136–63, and 7.

influential cartographic document for the subsequent representation of the European city.

Furthermore, in the sixteenth and seventeenth centuries numerous talented painters (Raphael, Michelangelo), architects (Bramante, the Sangallo family, Michele Sanmicheli), engravers, publishers and cartographers (Tempesta, Hondius), geographers and engineers (Vauban) occupied themselves with the design of military architecture and the depiction of battles and sieges. They were employed by the principal French and Spanish monarchs, such as François I, Philip II, Henri II, Henri IV, Louis XIII and Louis XIV, an entire series of Roman Popes, as well as Italian, German and Dutch princes and dukes, and oligarchies such as the United Provinces and Venice.

Throughout the sixteenth and seventeenth centuries these rulers and governments were busy in establishing clear domains of authority, defining, and even expanding, at each other's expense, the boundaries of their nascent nation-states. Thus this two-century-long period witnessed a great number of wars, characterised by movements of troops and the fortification of towns threatened by the growing militarized power of European sovereigns and facing the increased Ottoman threat from the east and the south.[3] The definition of statehood and its frontiers, the evolution of the administrative and financial apparatus of the new monarchies and the waging of war influenced one another mutually. These political, institutional and ideological conditions required the creation of standing armies and consequently the establishment of a powerful military order. War, central to the functional development of absolute monarchy—the form of government refined in the seventeenth century—and of the early modern state, transformed the appearance and functions of the early modern European city.

During this period of two centuries war was waged through battles and sieges. Naval battles and battles in the open field were rare, however (Legnano and Lepanto stand out), and most conflicts were resolved through sieges of towns, since the question was no longer to establish the greater ability and strength among equals (as in the morally defensible clashes inherited from the Middle Ages), but to demonstrate sovereignty.[4] Consequently, the great military conflicts of the sixteenth and seventeenth centuries were known by the names of the cities around which they were contested; Pavia, St. Quentin, Breda, La Rochelle are among the best known.[5] The objects of most sieges were strategically situated towns and cities whose inhabitants dominated financially and politically their surrounding territory, or controlled strategically the entry and passage into an important area. Fortification of these towns became paramount then for the defense of a sovereign's territory and claim to nationhood. Parallel with the interest and need for more efficient, resistant and affordable fortification, there developed a field of research in military strategy. Thus the practical requirements of war preparation were matched by an equiv-

3 See G.N. Clark, *War and Society in the Seventeenth Century* (Cambridge, 1958).

4 Simon Pepper, "The Meaning of the Renaissance Fortress," *Architectural Association Quarterly*, 2 (1973):21–28.

5 For the history of these sieges see Christopher Duffy, *Siege Warfare: the Fortress in the Early Modern World 1494–1660* (London, 1979).

alent theoretical counterpart focused upon the art, philosophy and science of war.

Because of its pervasive nature and ubiquity, the history of war has been the dominant focus of traditional political history, intent on analysing armed conflicts between cultural and linguistic groups. But war became a widespread object of study only during the Renaissance. There exists a large body of contemporary sixteenth-century publications in which the respective values and contributions of arms and letters are debated, the theme of "arma et litterae" in which the soldier is evaluated and compared with the scholar and the poet. Through this theme the identification between war and culture was soundly established; the result was the idealized association of learning and the ruling Prince in Renaissance political philosophy, together with the definition of a widely held ideal, the learned soldier.[6] The debate between the relative merits of arms and letters had an important precedent in Caesar who was both writer and warrior; and, even though it was assumed that women preferred military men to intellectuals, in educational treatises of the mid-seventeenth century literary study was considered a worthy occupation for a Prince at war.[7]

More recently the influence of war on the arts has been accorded new attention through the examination of the representation of military conflict in the oeuvre of talented Italian, French and German artists.[8] The literature of permanent fortification has been closely examined, predominantly by nineteenth-century Italian, French and German historians—many with a military background—who brought patriotic claims and local interests to their interpretations, vying to establish the claims to innovation of their co-national predecessors. Many of these publications can be characterized as bibliographic, some merely listing authors and titles in an attempt to estimate the quantity of the output and the possible numbers of editions. Thus we have, among others, the very reliable compilations by Cockle, Cosseron, Guarnieri, Maggiorotti, and Pohler.[9] In addition, there is the admirable work of Marini and Promis who resuscitated the writings of dozens of Italian military architects as part of their interest in establishing the hegemony of Italian architects in Renaissance fortification. The treatises on military architecture have been analysed in a masterly fashion by Horst de la Croix, who studied the Renaissance texts and their immediate predecessors, while Paolo Morachiello, Ulrich Schütte and Hartwig Neumann ranged over a wider chronology.[10] These architectural historians

---

6 John R. Hale, "Sixteenth-Century Explanations of War and Violence," *Past and Present*, 51 (1971):3–26.

7 Enea Piccolomini, *De duobus amantibus historia* (1444): "the clash of arms delights women more than the cleverness of learning"; Luigi Giuglaris, *La scuola della verità aperta a' Principi* (Venice, 1665), 10: "e del gran Carlo Emanuele [duke of Savoy] bastera dire, che anco nel giorno ch'espugnerò Trino havea studiato le sue hore."

8 John R. Hale, *Artists and Warfare in the Renaissance* (New Haven and London, 1990), takes a thematic rather than monographic approach.

9 *Cockle*, Cosseron de Villenoisy, *Essai historique sur la fortification* (Paris, 1869), Guarnieri, L. A. Maggiorotti, *L'opera del genio italiano all'estero* (Rome, anno XIV), Johannes Pohler, *Bibliotheca historico-militaris*, 4 vols.(Cassel 1887–95, Leipzig, 1899).

10 For complete citations see the *Abbreviations*.

and historians of printing have begun to consider the significance of this literary and graphic corpus as artistic entity and literary genre.

THE WRITERS OF TREATISES ON MILITARY ARCHITECTURE

The extensive military literature of the early modern period, then, illustrates vividly the interest taken in matters associated with war, and a significant segment of this literary output is focused upon the problems of military architecture—the permanent fortification of cities, towns and strategically located sites. These treatises on military architecture were written mostly during peacetime. Their authors demonstrate a great range of social origin, intellectual and literary preparation, experience in war as well as in their principal occupations. While early texts on war were generally authored by non-specialist writers, such as the philosopher Christine de Pisan, the papal abbreviator Roberto Valturio, and the political philosopher Niccolò Machiavelli,[11] the earliest innovations in Renaissance military architecture are owed to artists and architects such as Francesco di Giorgio Martini, whose architectural treatise remained unpublished until the nineteenth century, and Leonardo da Vinci.[12] The significance of military architecture had been acknowledged in Alberti's writings (c.1452) and by Filarete, whose treatise on architecture included a chapter on the form of the fortified ideal city;[13] fortifications were strongly visualized by Giuliano da Sangallo, Antonio da Sangallo, and Michelangelo, whose observations and drawings as well as realized fortifications were crucial in the development of an aesthetic of military architecture.[14] In addition to their research in fortification the authors of the treatises in this catalogue pursued a wide range of artistic, literary, scientific and editorial activities. Although a specialized focus on military architecture is detectable from c. 1575, during this two-century period there is, however, a shift of emphasis from an artistic to a scientific approach. This transformation is chronologically framed in this catalogue by Albrecht Dürer's treatise, the earliest illustrated book on military architecture, produced by a Renaissance artist known as one of the foremost engravers of western art (cat. no. 17–18), and Vauban's, the influential field marshal

11 Christine de Pisan, "Les Faicts d'armes et de chevalerie," Ms, BN Paris, c.1410; Valturio, *De re militari libri XII* (Verona 1472); Machiavelli, *Dell'arte della guerra* (Florence 1521).

12 Francesco di Giorgio, "Trattato dell'architettura civile e militare" (dated variously between 1482 and 1495), ed. Corrado Maltese (Milan, 1967); Leonardo, "Codice Atlantico," (drawn between 1479 and 1513); for an examination of his drawings for fortresses see Pietro C. Marani, *L'architettura fortificata di Leonardo da Vinci* (Florence, 1985).

13 Alberti, *De re aedificatoria* (Florence, 1485); Filarete "Trattato dell'architettura," c.1461, BN, Florence (still largely manuscript).

14 For the Sangallo see C. Huelsen, *Il libro di Giuliano da Sangallo* (Leipzig, 1910), Giancarlo Severini, *Architetture militari di Giuliano da Sangallo* (Pisa, 1970), *Antonio da Sangallo il Giovane; La vita e l'opera*, Atti del XXII Congresso di Storia dell'Architettura (Rome, 1986), and the forthcoming catalogue of drawings by the Sangallo at the Uffizi (Florence) with the sections on fortification by Nicholas Adams and Simon Pepper; for Michelangelo's contributions see James Ackerman, *The Architecture of Michelangelo* (London, 1961), ch. 5, and Vincent Scully, "Michelangelo's Fortification Drawings: a Study in the Reflex Diagonal," *Perspecta*, 1 (1952):38–45.

and strategist of Louis XIV (cat. no. 70–72).¹⁵

Talented architects such as Pietro Cataneo (cat. no. 8) and Vincenzo Scamozzi (cat. no. 54) included chapters on fortification in their treatises, although their model and mentor Palladio was contemptuous of military architecture—believing that the only fortification a city needs is the strength and devotion of its citizens. But war and fortification were evidently on his mind: Palladio studied Polybius and wrote a preface to Caesar's *Commentaries*, dedicated the third book of his treatise to Emanuele Filiberto, duke of Savoy—because only he, among contemporary princes, resembled ancient Roman military heroes—and in the preface to his treatise on architecture promised to discuss the method of fortifying cities.¹⁶ His moral qualms, voiced elsewhere earlier in the century during the debates that preceded fortification,¹⁷ were largely abandoned by late Renaissance and Baroque political philosophers and scientists who contributed to the literature on war.¹⁸

Of the authors represented in this catalogue, Cellarius, Tartaglia, Marolois and Stevin were mathematicians; Hondius, Dubreuil, de Fer and Bodenehr were cartographers, printers and publishers; Fournier, Tacquett and Rossetti were Jesuits involved in teaching; Maggi, Gualdo-Priorato, Ozanam and Sturm were above all historians; while Coehoorn and Vauban were the most innovative military strategists of the seventeenth century. Most of the other authors were principally artillery officers, such as Pagan, de Ville, Tetti, Alghisi, Busca, Croce, Groote, Sardi, Tensini, Floriani, or engineers, such as Zanchi, Cataneo, Speckle, Perret, Errard, Freitag, Goldmann, Dogen and Capra, whose writings were largely based upon their personal experience in war.

The principal printing center for military treatises was Venice,¹⁹ whose earlier, sixteenth-century hegemony was followed and then replaced in the seventeenth century by Paris, with London, Amsterdam, Frankfurt and The Hague as close seconds (although both London and Frankfurt, and later Amsterdam, were important in publishing works in translation rather than original books). Other important treatises were printed in various north-Italian towns such as

---

15 The classic monograph on Dürer is Erwin Panofsky, *Albrecht Dürer* (Princeton, 1955), but his fortifications treatise is analysed in William Waetzoldt, *Dürers Befestigungslehre* (Nuremberg, 1916), and *Dürer and his Times* (London, 1950), 220–26; see also Martin Biddle's introduction to *Etliche Underricht / zu Befestigung der Stett / Schloss / und Flecken* (1527; rpt. New York 1972); for Vauban see Blomfield, Pierre Lazard, *Vauban, 1633–1707* (Paris, 1934), and *Parent and Verroust*.

16 *I quattro libri* (Venice, 1570); see also John R. Hale, "Andrea Palladio, Polybius and Julius Caesar," in his *Renaissance Studies* (London, 1983), 471–86.

17 John R. Hale, "To Fortify or not to Fortify? Machiavelli's Contribution to a Renaissance Debate," in *Essays in Honor of J. H. Whitfield* (London, 1975), 99–119.

18 A political philosopher such as Giovanni

Botero, active at the end of the sixteenth century, wrote not only about the absolutist form of government desirable for post-tridentine Italian states in his *Della ragione di Stato* (Venice, 1589), but also composed a treatise on fortification for the education of the Princes of Savoy (included in later editions of the *Ragione*).

19 See John R. Hale, "Printing and Military Culture of Renaissance Venice," *Medievalia et Humanistica*, Studies in Medieval and Renaissance Culture, ed. Clogan (Cambridge, 1977), n. s. 8:21–62.

Turin, Milan, Florence, Vicenza, Brescia and Macerata, in Rotterdam, Leiden and Antwerp, throughout Germany—in Osnabruck, Strasbourg, Augsburg and Nuremberg—and in Lyons.

As the instrumental theoretical texts of what later was labelled the "international architetural style" of the sixteenth and the seventeenth centuries,[20] the treatises on military architecture were in great demand across national and linguistic boundaries, not unlike scientific papers today. Although the treatises by Dürer, Lanteri, Dilich, Goldmann, Dögen and Tacquett were rendered into Latin—still the lingua franca of the intellectual and governmental communities—subsequent treatises, especially in the seventeenth century, were translated mostly into other vernacular languages, thus demonstrating the intention of the publisher to reach a wider readership. Valla's book went into nine Italian editions between 1524 and 1550, with one French edition in 1554; Tartaglia's *Nova Scientia* came out five times between 1537 and 1606; Zanchi's treatise was printed in 1550, 1554, 1556, 1560, and 1601 in Venice and Lyons; Gerolamo Cataneo's treatise was published in five Italian and two French editions between 1564 and 1608.

This process sped up in the seventeenth century. Perret's graphically spectacular publication came out in 1601 in Paris, in 1602 in Frankfurt, in Oppenheim in 1613 and again in Paris in 1620; Errard's treatise was published in Paris in 1600, 1604, 1620 and 1622, and then in Frankfurt in 1604 and 1617; Marolois's French and German editions emerged in Holland in 1615, 1627, 1628, 1637, 1638 and 1651. De Ville's *Fortifications* saw seven editions between 1628 and 1672, and Freitag's six between 1631 and 1665, while Dögen's treatise was published simultaneously in Latin, French and German. Ruse's treatise was printed twice in German and soon afterwards in Dutch and English; Mallet's "history" of fortification was published in Paris and The Hague in 1671, 1685, 1691, 1696, and in German in Amsterdam in 1672. But Coehoorn and Vauban's publishing success far outshone that of previous treatise writers. Coehoorn was published in Dutch, French, English, German and Russian; while Vauban's ideas were plagiarised in Paris, Amsterdam and London, and the rendering of Vauban in German by the celebrated polymath Sturm in the *Architettura militaris* emerged in Nuremberg in 1702, 1719, 1720 and 1735.

THE INTENDED READERS OF TREATISES ON MILITARY ARCHITECTURE

But who were the consumers of this highly specialized and technically detailed literary corpus? The readership of this large body of works can be gleaned partly from the dedications and prefaces provided by many of the authors, the large number of publication centers for military architecture, and the numerous editions and translations of the individual treatises. The most privileged readers were the immediate employers or patrons of the military theorist to whom the innovations of the latter were offered. Occasionally, the most highly placed patrons were offered the exclusive reading, appreciation, and profit of the military architect's theoretical pre-

---

20 Stanislas von Moos, *Turm und Bollwerk; Beiträge zu einer politischen Ikonographie der italienischen Renaissancearchitektur* (Zürich, 1974), ch. 4.

cepts and practical realizations. Thus, for example, Vauban consigned his work to secrecy by refusing its publication. Furthermore, he himself devised the ritual manner in which the manuscripts of his instructions for the defense and attack of fortresses should be handed down among the army commanders through the direct mediation of the king.[21] Eventually, the multiplied manuscripts found an unguarded outlet and were offered in plagiarized and fragmentary form—it is the case of the Vaubanian treatises in this catalogue—to a public avid to know the military secrets of the French nation. As we have seen above, with equal awareness of the value of his innovations, Leonardo offered the Duke of Milan the possibility of exclusive use of the ingenious war machines that he had invented. The designs implemented by a military architect were often considered the intellectual and certainly the military property of the patron. Floriani, for instance, had to wrest from Pope Urban VIII the permission to publish his treatise on fortification, a right contested by the Pope because it entailed the unveiling of state secrets.[22]

The dedicatory notices of the treatises offer further insight into the readership envisioned or postulated by each author, while exposing both the author's intellectual ambitions and his social aspirations. Dürer dedicated his treatises to Ferdinand I, the King of Hungary and Bohemia, brother of Emperor Charles I and grandson of Emperor Maximilian I. Dürer was the beneficiary of an imperial pension from Maximilian and counted among his friends Ferdinand's superintendant of fortifications.[23] With the exception of two treatises—those by Stevin and Capra, which seem to be dedicated to an intimate friend of the author or the publisher[24]—this group of military treatises is dedicated to Holy Roman Emperors, kings of western and central Europe, German, Italian and Dutch princes, and aristocrats known for their military interests and involvement in governmental bodies such as the Venetian Senate, the English Privy Council, or the French war ministry. But the largest numbers of treatises were dedicated to Italian and German Dukes, Doges and Archdukes, rulers in their own right or potential heirs of royal and imperial titles.

It can be deduced then from the dedicatees, that the fortification treatises were intended for chiefs of great armies, military leaders, condottieri, diplomats, political personalities, and the larger numbers of the nobility. They were also intended as a dialogue between the members of the growing community of artillery officers, engineers, strategists and mathematicians involved in warlike activities, such as the construction of fortification, the management of sieges and the organization of armies. Furthermore, since wars were the dominant concern and among the most widely practised activities of the sixteenth and seventeenth centuries, this literature no doubt also found a large general audience.

---

21 Joan DeJean, *Literary Fortifications* (Princeton, 1984), 34–36.

22 Biblioteca Apostolica Vaticana, Barb. Lat. 7788.

23 Alexander von Reitzenstein, "Etliche Unterricht . . . ; Albrecht Dürer's Befestigungslehre," *Albrecht Dürer's Umwelt* (Nuremberg, 1971).

24 See cat. no. 6, 59.

The sixteenth- and seventeenth-century treatises on military architecture were dedicated, then, to the most powerful members of the aristocratic elite. In his study of some of these sixteenth-century treatises Sir John Hale has convincingly demonstrated, through his analysis of their title-pages, that the dedications—supported by the composition of engraved frontispieces—attempt to uplift the social and intellectual standing of the author by this association with a famed military leader or diplomat.[25] Thus the production of military treatises by the practitioners of military architecture can be seen as an effort to endow the profession with a theory (since the possession of a theory, even flawed, lifts it from a mechanical to a liberal art through intellectual abstraction of general rules) and to ennoble it by giving a measure of responsibility for cultural and political life. The attempt is then to endow a practical profession, only marginally acknowledged in the fifteenth and the early sixteenth century, with an intellectual context and, by associating it with the interests of the powerful, to elevate the profession's social standing as well.

The Intellectual Context for Military Architecture

Simultaneously with the definition of the roles that military architecture and engineering were to play in Renaissance and Baroque war, accomplished in the sixteenth century through the great technical innovation of the polygonal bastion, the literature of this branch of architecture began to integrate itself with the general interests of the period and to mirror its intellectual concerns. These intellectual concerns included the adoption of principles from Renaissance art and aesthetic theory, especially the concepts of perspective and magnificence, the search for ancient parallels (the fundamental Renaissance polemical ethos in rhetoric and philosophy), the definition of social order through the form of the ideal city (pioneered by architects such as Filarete and Leonardo and subsequently by political philosophers like More, Botero and Campanella),[26] and the application of a mathematically-based scientific approach closely attuned to the work of seventeenth-century philosophers of science.

The adoption of aesthetic principles to the design of bastioned fortification was the fundamental contribution of fifteenth- and early sixteenth-century Italian architects, and must be seen as an extension of the interests in classical art and architecture that characterize the Renaissance. Even though most of the precedents and models available to architects dated from the Middle Ages, in Rome the ancient walls of the city had been largely preserved (and even restored, though for antiquarian rather than strategic reasons). Thus by the 1510s Renaissance architects had to come to terms with the ancient Roman towered brick walls whose monumentality of conception, grandiose proportions and masterly construction were a worthy counterpart of the triumphal arches, baths and theaters that littered

---

25 "The Argument of Some Military Title Pages of the Renaissance," *The Newberry Library Bulletin*, 6 (1964):4:91–102.

26 Thomas More, *Utopia* (1516), Giovanni Botero *Delle cause della grandezza delle città* (Rome 1588), Tommaso Campanella, *Civitas Solis* (1602)

Rome and inspired the architectural revival of the Quattrocento.

The idea of beauty coupled with awesomeness in military architecture may prompt a smile among those (military) historians who focus on its utilitarian aspect, but the bastioned trace was designed to be attractive by its fifteenth and sixteenth-century architects, and was expected to be beautiful by its commissioners.[27] Later, in the seventeenth century, awesome fortifications were considered beautiful if large and carefully built. Thus Turin's fortifications, reputed as among the most powerful in Italy by 1630, were described as *bellissime* by awed visitors, while Louis XIV did not mind how much a fortification cost as long as the result was a *belle place*.[28] No one can deny the visual impact of Francesco de Giorgio Martini's fortifications for the Duke of Urbino nor the visual strength of Giuliano da Sangallo's Nettuno fortress or Antonio da Sangallo's Ardeatina bastion in Rome.[29] The bastioned fortification was a significant addition to the architectural vocabulary of the High Renaissance, whose most talented architects practised both civil and military architecture. Thus for the Farnese palace at Caprarola the architects Sangallo, Vignola and Paciotto adopted the bastion at the angles of the pentagonal outline of the building.[30] While the circular court of the pentagonal palace emulated the famed cistern well of the fortress at Orvieto, it also echoed the court of the most important Spanish residence, the palace of Charles V at Granada's Alhambra. The gates that the architect Sanmicheli built in Verona were intended to be both physically threatening and visually striking, recalling Rome's ancient gates and by implication its imperial power.[31] Paralleling the search for the fundamental architectural element—the design of the five orders of architecture which interested so many of the best Italian and French architects—military architects developed a "military" order: the column whose shaft was strengthened with brawny banding strips influenced perhaps by Philibert de l'Orme's proposals for the "French" order.[32] The collusion of beauty and functionalism had already been proposed in Vitruvius' treatise on architecture, whose tenth book was dedicated to fortification, siege engines and other aspects of war. The other great ancient Roman source for military art was the fourth-century writer Vegetius, who became the most copied military author after Julius Caesar.[33]

---

27 See John R. Hale, *Renaissance Fortification: Art or Engineering* (London, 1977).

28 *Il conte Fulvio Testi alla corte di Torino negli anni 1628 e 1635*, ed. D. Perrero (Milan, 1865), 126; Joan DeJean, op. cit., 21–23.

29 For Francesco di Giorgio see Francesco Paolo Fiore, *Città e macchine*; for the contributions of Antonio da Sangallo and his uncles to military architecture see the section "L'architettura delle fortificazioni," in *Antonio da Sangallo il Giovane; La vita e l'opera*, Atti del XXII Congresso di Storia dell'Architettura (Rome, 1986).

30 Loren Partridge, "Vignola and the Villa Farnese at Caprarola," *Art Bulletin*, 52 (1970):81–87, and Italo Faldi, *Il Palazzo Farnese di Caprarola* (Turin, 1981).

31 See *Michele Sanmichele, 1484–1559; Studi raccolti dall'accademia di agricoltura, scienze e lettere di Verona* (Verona, 1960), and Earl Rosenthal, *The Palace of Charles V at Alhambra* (Chicago, 1985); E. Langskioeld, *Michele Sanmichele, the Architect of Verona* (Uppsala, 1938).

32 See Anthony Blunt, *Philibert de l'Orme* (London, 1958), 118–22.

33 Flavius Vegetius Renatus, *Epitoma rei militaris*; Marcus Vitruvius Pollio, *De architectura*.

Vitruvius' and Vegetius' works, available to fifteenth-century readers through numerous manuscript copies, were eventually published and continued to be important references for the entire period under consideration here. Their contributions, augmented by those of lesser Roman writers, provided the foundation and precedents upon which the early modern military treatises were modelled.[34]

This modelling consisted—at least in the early treatises—not only in direct borrowing but in providing a veneer of erudition and learning and thus endowing military treatises with the prestige of classical literature. For example, although Palladio in his treatise on architecture did not discuss the fortification of the city, he had nonetheless carefully studied both the ruins of ancient Rome's fortification and Caesar's literary remains (of which he published an edition), as well as illustrating Daniele Barbaro's edition of Vitruvius.[35] Even though by the end of the sixteenth century the technique of defense had evolved in response to the revolutionary developments in attack strategy brought about by the use of firearms, some of the most learned and talented military theorists (as we have seen above), continued to have their works translated into Latin, the language of Renaissance humanistic scholarship as well as seventeenth-century science, whose use was based upon an unquestioned belief in the superiority of classical learning. In addition, many authors prefaced their works with a discussion based upon ancient classical writings on war, siege, fortification and the management of the army.

But the principal theme of early modern military theory to be influenced by ancient sources was the organization of the army. This consisted of the training, movement and the lodging of large groups of soldiers.[36] While these concerns are somewhat beside the interests of permanent fortification, they have nonetheless an architectural and urbanistic counterpart in the problem posed by the encampment. Often the requirement for rationalizing the military encampment was translated into an opportunity to design an ideal town.

The earliest Renaissance architectural theorists concerned themselves intensely with the conception and planning of the ideal city. Francesco di Giorgio Martini devoted some of his most elaborate drawings to this problem, while Filarete and Leonardo da Vinci not only provided designs, but added their own conceptions of social order which were expressed through the architectural form and the general layout of their proposed towns. Based upon aesthetic principles simultaneously developed in painting and in architecture, their conceptions were centralised plans with radial or orthogonal grid streets, with centrally located piazzas, and with gates and bastions on axes with one another. Their compositions were based upon symmetry, hierarchy and harmony of the individual parts with one another.

These ideal city planning principles were the counterpart of the major archi-

---

34 For a succint resumé of Vitruvius's literary influence see *Wiebenson*, Introduction.

35 Andrea Palladio, ed., *I commentari di C. Giulio Cesare* (Venice, 1575); Daniele Barbaro, ed., *I dieci libri dell'architettura di M. Vitruvio tradutti et commentati* (Venice, 1556).

36 Valla, cat. no. 68, Roberto Valturio, *De re militari* (Verona, 1472), Leonhardt Fronsperger, *Kaiserlichen Kriegsbuch* (Frankfurt, 1564).

tectural interest of the fifteenth century, that is, the centrally planned church which reflected the humanistic credo that man stood at the center of the universe for which he served as proportional system; but they also showed the autocratic controlling tendencies of contemporary governments. Their tendency to unified control became the governing force of several of the most influential west-European countries by the end of the sixteenth century, and concretized itself in the absolute monarchies of the seventeenth century. It has been said that the ideal city of the fifteenth- and sixteenth-century utopian architects (Leonardo, Francesco di Giorgio) and philosophers (Erasmus) became the ideal fortress of the seventeenth-century military architects.[37]

The ideal city was conceived out of the desire for classically-founded planning principles. These consisted of a central governing body which could implement preconceived, that is, theoretically conceived urban form which could then be applied to any given site or condition. The variety implicit in towns which had grown over time was sacrificed in the projects of Francesco and Leonardo in order to achieve a coherent aesthetic order based upon the abstract principles of symmetry and proportion. Filarete, on the other hand, planned his town and the social order that would inhabit it in a typological manner based on the concept of 'varietas,' designing each kind of building in its turn, where the appropriate form was proposed for each of the desired functions. This was Dürer's approach as well, although for his overall plan he relied on the most basic and widely adopted plan of the ancient Roman castrum: the orthogonal street grid modelled on the established encampment of the Roman army and adopted for the settlements of its veterans (cat. no. 18).[38] Although radial plans were also valued and appeared early on in the designs of Fra Giocondo and in the layout of Palmanova (cat. no. 33),[39] the orthogonal street grid was more popular in civil and military architectural treatises: the plans by Dürer, Cataneo, and Scamozzi, among the authors in this catalogue, come readily to mind. This neo-platonic polygon—the square—was added to the circle, the other Euclidian form favored by Renaissance artists, and formed the basis of early Renaissance studies in geometry that provided the theoretical foundation for the more complex polygons with which military architects experimented after the 1550s in their proposals for efficient and beautiful fortification enclosures.

It is in their determined emphasis on a mathematical and scientific approach that the works of seventeenth-century theorists of military architecture distinguish themselves from their immediate predecessors. Although "ars sine scientia nihil est"—a medieval tenet still current in the Renaissance—might be updated as "art cannot exist without knowledge" in fact it was understood at the time as "craft (art being the customary way in which things were done) is

---

37 Pierre Francastel, "Paris et la création urbaine en Europe au XVII siècle," in *L'Urbanisme de Paris et l'Europe 1600–1680*, ed. Pierre Francastel (Paris, 1969), 9–37.

38 Dürer, cat. no. 17.

39 Horst de la Croix, "Palmanova, a Study in Sixteenth-Century Urbanism," *Saggi e Memorie*, 5 (1962):23–41.

not sufficient without a set of guiding principles." Since fortification against cannon became fully formulated only about 1550, the guiding principles used for the design of bastions were somewhat loosely assembled. Once the technological revolution brought about by the adoption of firearms was fully integrated as the way of making war and theorists no longer debated the value of fortifications, they began to concentrate upon methods of design that could be readily applied to the construction of bastioned fortifications, and upon the theories that were to be taught to aspiring military architects. The technological revolution included the improvement of field cannon, a more accurate understanding of ballistics (Tartaglia's treatise provided a foundation) and greater ease with vaulted construction.[40] Although from the very beginning of this literary type, c. 1550, each military theorist who wrote a treatise concentrated upon formulating a series of coherent principles that one could learn and then put into practice as a methodical system of fortification, a rationally consistent discourse was only developed at the turn of the seventeenth century. In this context the main interest was in the "system" which integrated the innovations and contributions made by the individual theorists-engineers. While at least two of the sixteenth-century treatises in this catalogue propose research in the principles of geometry as the basis for learning the design of the bastioned fortification trace (Dürer and Lanteri), among the seventeenth-century treatises mathematicians dominate the field. Similarly, while in the sixteenth-century treatises surveying is often the foundation for learning to lay out the fortification trace (Cataneo, Tetti), surveying becomes the implicit foundation, almost a commonplace, in seventeenth-century works on military architecture. Both surveying and studies in Euclidian geometry are areas of interest shared by civil and military architects. What is new in the seventeenth-century works is the concern to consider all possibilities of site and of size, and to abstract from them a series of precepts that can be universally applied.

As the interest in aesthetic excellence, ancient models and contributions to the definition of the social order continued unabated among military architects, the impulse towards scientific veracity was reinforced by developments in seventeenth-century research and theoretical innovations in science and philosophy. Thus seventeenth-century theorists benefited from the contributions of such great scientists as Galileo and Descartes, both of whom brought to bear upon military architecture the research methods they had employed respectively in astronomical and physical studies.

The polemic at the core of these treatises, that is the competition between a theoretical and a historical approach, was exacerbated after 1600. What seventeenth-century writers on military architecture questioned was whether the theories should reflect the accumulated lessons of individual experience—their own and that of their predecessors (including the ancient Roman writers)—or whether they should concentrate on how one might build fortifications un-

---

40 For Tartaglia see Alexandre Koyré, "La dynamique de Nicolo Tartaglia," in his *Études d'histoire de la penseé scientifique* (Paris, 1973):117–39.

der theoretical, that is, ideal, circumstances. They were responding to a revolution of mental attitudes that has been described as "reducible to two fundamental and closely connected actions: the destruction of the Cosmos and the geometrization of space, in which the Aristotelian conception of space is replaced by Euclidian geometry—an essentially infinite and homogenous extension—considered as identical with the real space of the world."[41] Characteristic of this revolution were the secularization of consciousness, a turning away from transcendent goals to immanent aims, the discovery of the subjectivity of human consciousness, and a shift in the relationship between *theoria* and *praxis*.[42] The world as a finite, closed and hierarchically ordered whole was replaced by an indefinite and infinite universe where the world of value was divorced from the world of facts. Thus military architects no longer needed to consider the moral implications of fortification construction but concentrated—unburdened—upon the improvement of their war strategies.

The definitive welding of military architecture with mathematics and geometry occurred through the scientific approach adopted in the seventeenth century. Thus even an engineer (a builder of machines) could be capable of scientific discourse, a discourse of principles and origins, because he was aware of causes. It was Lorini who guided military engineering to the threshold of modern science; Errard transformed the mechanical art of fortification into a "perfectly demonstrable art," and therefore scientific. Perret identified the science of fortification with geometric construction, while Freitag's treatise is one in a long chain of works on geometry and trigonometry in which the author believes that everything, even the most pragmatic aspects of fortification, can be calculated in advance.[43] This notion of accountability was supported by Marolois, who believed in the dominance of theoretical design, and amply realized in Vauban's *cahiers de charge*.[44]

This scientific approach based on mathematics was promoted by the Jesuits, who from 1551 taught the subject at their Roman headquarters and eventually founded chairs in mathematics in France. Additional chairs were established by French royal decree in order to train officers and administrators. But as late as 1627 classes in mathematics were offered only at the Paris Jesuit College and at La Flèche in Normandy, while the students were mostly young nobles preparing for a military career.[45] The Jesuits' monopoly of education in Catholic countries allowed them to dominate also the teaching of mathematics; both conditions explain why so many of the treatises were written by Jesuits—familiar only with the theoretical aspects of war and fortification.

"Space confused with geometry" and "time confused with numerical sequence" might be another way of interpreting the seventeenth-century mental attitudes which extended naturally into

---

41 Alexandre Koyré, *From the Closed World to the Infinite Universe* (New York, 1958), vi.

42 ibid., 4.

43 Alessandro Biral and Paolo Morachiello, "Filosofo, soldato, politecnico," in *Manno*, 62–65.

44 ibid., 51–57.

45 See François de Dainville, *L'Education des jésuites, XVIe-XVIIIe siècles* (Paris, 1978), 323–37.

military architectural theory.[46] Conception of nature as either God-given geometrical order and mathematical reality or as a wilderness, a labyrinth to be navigated through the use of method, encouraged scientific and theoretical research.[47] For Descartes, a student at La Flèche, everything seemed to be a science, his universe was a "grid of arithmetical relations." (Although scholasticism had named "mathematics" all disciplines which examined order or measure, no matter what the object of examination was, in the seventeenth century the term acquired greater specificity.) Having established reasoning as the sole activity reserved exclusively to humanity, Descartes pointed the way towards continuous research and examination, preventing scientists from merely accepting each other's results. In military architecture this translated into numerous individual works competing for attention. Thus, while the early treatises of cannon fortification (published in the sixteenth century) responded to a technical revolution that threatened the existence of social arrangements and organizations, the treatises published in the seventeenth century confronted different intellectual conditions which made new demands upon their innovative and pedagogical abilities.

The Treatises

The theorists of military architecture, then, contributed to the burgeoning literature on subjects related to war. In addition to addressing the humanistic themes discussed above—the interest in aesthetic principles and the revival of antiquity during the fifteenth and sixteenth centuries, the adoption of a mathematical approach in the seventeenth century, and the contribution to the definition of a social order—the treatises on military architecture stressed a series of problems which became the commonplaces internal to the discourse on fortification. Most of the treatises presented here offer design solutions, illustrating or evaluating what the nineteenth-century historian Luigi Marini has called permanent fortification.[48] The principal concerns of this internal discourse on permanent fortification are the mathematical conception, the form, the representation, the development and the refinement of the bastioned trace, strategies for the defense and attack of fortified cities, and the location and use of the citadel, the main fortification element of a city.

The principal themes of the fortification treatises published before 1600 in Italy have been presented by Horst de la Croix, who maintained that although the influence of Italian engineers declined in the seventeenth century their earlier contributions provided the basic concepts for bastioned fortification.[49] The Italian innovations in defensive architecture were further explored by northern engineers whose ideas helped develop "national" schools of military architecture,[50] while the two significant transformations discussed above—the mental shift towards the "geometrization" of space, and the transfiguration of

---

46 Quoted from Arthur Koestler, in Paolo Rossi, *La scienza e la filosofia dei moderni: aspetti della rivoluzione scientifica* (Turin, 1989), 90.

47 Rossi, ibid., 91 and *passim*.

48 *Marini*, vol. 1, 4–36.

49 *De la Croix*, 49–50.

50 Bourdin, cat. no. 4, p. 66: "ordres de fortification".

the ideal city into the ideal fortress—altered the content and structure of the treatise, turning it into a fundamental work of reference and a continuously updated scientific text. The common purposes of the military treatise writers—to carve out a professional standing, to establish claims to innovations and contribute to the history of fortification—are dominated by the intention to instruct. The authors attempt to educate the reader in urban fortification by conveying the elementary principles of military architecture. These were derived primarily from personal experience, consequently abstracted into a series of mathematical precepts.

Geometry

Galileo's *Breve istruzione all'architettura militare* (dated 1593), and his slightly longer *Trattato di fortificazione*, provide a good example of the widespread interest in fortification and of the scientific method employed in the composition of the fortification treatise. Both of these works, discussed by Catherine Wilkinson,[51] begin with a lesson of geometry intended to teach the projecting of the right angle, the subdivision of the angle and the description of the regular polygon. Having reiterated the reason for fortification (to resist with a small number of soldiers a large attacking army), Galileo lists the offensive instruments against which the architect has to fortify, such as the battery and mines, while siege and assault remain the domain of the military commander.

[51] "Renaissance Treatises on Military Architecture and the Science of Mechanics," in *Les Traités d'architecture de la Renaissance*, Jean Guillaume, ed. (Paris, 1988), 467–76.

Summarizing sixteenth-century fortification philosophy, he proposes *flanking* as the fundamental defense, where every part of the fortification is visible from another part, and logically points out that it is best for this flanking defense that all its lines be equal. With these two precepts he provides an authoritative resume of the fortification strategy promoted in many works which are contemporary with his, principally those by Francesco de' Marchi (published only in 1599) and Jean Errard (1600).

In the *Trattato* Galileo teaches step by step the geometrical layout of the bastions, going beyond the brief listing of the elements of the bastioned trace provided in the *Breve istruzione*. His definitions of *plan* and *section* helped to establish further these modes of representation in two dimensions (not by chance are they the most abstract), while the explanation of scale rendered these drawings reliable documents. The greater part of both treatises is focused upon the description of the fortified trace's elemental parts, which include beside the polygonal bastion the platform, the cavalier (placed between or above bastions to defend the countryside around the fortification), and various outworks such as "scissors" (two half-bastions connected by a curtain wall), and "teeth" (made of flanks and curtain walls and used to fortify irregular hilly sites). Like other military writers Galileo provides optimal dimensions for all parts of the fortress, discusses the various kinds of sites an engineer might be asked to fortify, and provides step by step guidance for the laying out of the construction site and the construction sequence. This then is the structure of the fortification treatise composed by one of the major scientists of the seventeenth century. Without experience in siege, he was able and interested to compose a theoretical text for the instruction of future engineers; he provided a scientifically streamlined model for seventeenth-century authors of military treatises. Others wrote about the geometrical foundations of military architecture. Marolois' treatise on fortification was only the third part of a larger work, the *Opera mathematica*, which dealt also with geometry, trigonometry and perspective. The treatise by Henry Hondius, Marolois's friend and publisher, also contains an appendix entitled *Breve instruction des regles de la geometrie; fort utile et necessaire a la perspective et architecture militaire, ou fortification*.

Some theorists avowed preferences for certain regular polygons. Thus Menno, Baron of Coehoorn—considered the "Dutch Vauban"—writes at length about the "royal hexagon," which is also Sardi's favored fortification form (cat. no. 52). Errard's book on fortification begins with the design of the hexagon (according to him the first regular polygon that can be easily fortified), but then goes on to demonstrate the composition of polygons with up to 24 sides. This obsessiveness is implicitly criticized by Dögen, who stops his own demonstrations at the octagonal fortress, pointing out that no one was building greater regular fortifications (the nine-sided Palmanova near Venice had been the only exception). Subsequently, Pagan considered his own pentagonal and dodecagonal fortifications the most powerful, while Antoine de Ville—his most talented competitor—devotes many pages to the analysis of actual pentagonal fortresses, in Turin,

Antwerp, Casale Monferrato and Rome. Vauban's treatise, written to defend the superiority of his ideas, begins with a chapter on the principles of geometry necessary for the design of fortification. Preceding his observations on regular geometrical forms, Malthus provides an entire text on "practical" geometry, instructing on measurements of heights and distances, the use of the compass, and the principles of trigonometry. His fortification treatise goes beyond geometry and mathematics, and discusses artillery methods adapted from fireworks, thus transposing knowledge employed for recreational purposes to war activities. The calculations relating to artillery (the trajectory and the shooting range of various cannons and other firearms was a fundamental consideration for the dimensioning of the fortress's elements), anchored in geometry, trigonometry as well as ballistics, thus become principles upon which fortification design theory is founded. Bitanvieu, Ruggiero, Ruse and Capra, writing in the last third of the seventeenth century, begin their treatises with a definition of the fundamentals of Euclidian geometry (point, line, angle) and description of regular polygons, showing that the fortification treatise was meant as a textbook and how-to manual for would-be military architects and engineers.

REPRESENTATION

The most influential subject of these treatises for the history of cities is the representation of fortifications. While there are suggestions among the sixteenth-century texts that a drawing of a fortress might clarify in advance the problem implicit in a design solution,[52] there seems to have been no systematic method for drawing parts or entire fortifications until the seventeenth century when representation became closely linked to an understanding of the fortress's parts.

The interest in representation linked military architecture to civil architecture. The use of plan, section and elevation had been promoted in the sixteenth century by Raphael, for the historical documentation of Rome's quickly disappearing ancient monuments,[53] and by Daniele Barbaro, in his discussion of Vitruvius' *De Architectura* (1556). The three means of representation, *orthographia*, *ichnographia* and *scaenographia* were first outlined by Vitruvius, but Barbaro substituted *sciagraphia* for the third term, rendered in Italian "profilo" and thus derived the plan, elevation and profile, a "trinity of commensurable graphic representation more in keeping with his own sense of disposition."[54] Significantly, this system of representation was not consistently applied either in the design or documentary stages of a building until its espousal in these treatises on military architecture. One-point perspective had been first accurately constructed by the architect Brunelleschi in the first quarter of the fifteenth century but remained scientifically inexplicable—despite Alberti's analysis[55]—until the discovery of the vanishing point by the mathematician Guidobaldo del Monte, friend of both Tartaglia and Galileo.[56] Galileo defined the plan and section at the end of the sixteenth century and in 1625 Hondius listed three means of representation—ichnographic, orthographic and scenographic—but it was Dögen (1648) who defined clearly the three means of graphic representation that corresponded to these terms: the horizontal section or plan, the vertical section or profile, and the perspective, axonometric or bird's eye view; he used these means of representation in order to explain the terminology of military architecture. This scientific definition of plan, section and perspective view was of fundamental importance for the ability to visualize fortifications before building them in order to check dimensions, accuracy and strategical planning, as well as to present a proposed design to a patron. The plan and section provided in addition a very accurate method of documenting existing fortifications and aided in reconstruction and restoration efforts.

Treatises on military architecture were often preceded, as we have seen, by an introduction to basic geometry, paralleled by similar introductions to treatises on civil architecture. The contribution of the military theorists consists in the readiness with which they embraced discoveries made both in ar-

---

52 Alghisi, cat. no. 1, book 1.

53 In a memorandum, composed with the aid of Castiglione, to Pope Leo X.

54 See David Rosand, *Painting in Cinquecento Venice: Titian, Veronese, Tintoretto* (New Haven, 1982), 178; and entries on Barbaro by the same author in *Wiebenson*.

55 For Brunelleschi's contributions to perspective see Richard Krautheimer and Trude Krautheimer-Hess, *Lorenzo Ghiberti* (Princeton, 1956), 234–48; *Della Pittura*, (1436).

56 *Perspetivae libri sex* (Pesaro, 1600); see entry by Paul Breman in *Wiebenson*, III-B-12, and the discussion by Catherine Wilkinson. op. cit., 469–70.

chitecture and mathematics, their quick understanding and acceptance of scientific methods, and their adoption of scientific terminology as ornament for their treatises. In addition, the acceptance of measured drawings and surveying methods represented a close link with another endeavor, that of cartography. Hondius is an important example in this context because while he wrote on military architecture and on perspective, he was also a publisher and cartographer in his own right.

Most importantly military architecture and its literature altered the image of the city beginning with the earliest published treatises. Through these works military theorists disseminated the image of the European city with a geometrically perfect lay-out surrounded and defined by impeccably sharp fortified defenses. This sharp definition endowed the representations of fortifications with an immediacy that was palpably stronger than the realities of the actual city, whose walls could never be perceived as a whole, except in a cartographic illustration. The visual strength achieved through the manipulation of perspective made the fortifications seem even more fearful than they were in actuality.

In producing these powerful images military architects were aided by the quickly developing techniques of Renaissance cartography which they wholeheartedly adopted and advanced. Many military writers had close connections with cartographers or were actually cartographers themselves, and the earliest accurately surveyed manuscript plans of cities known to us were made for military purposes.[57] Thus, beside adopting and refining methods of representation borrowed from civil architecture, the illustrations of the military treatises benefited from the results of accurate surveying, an endeavor that was greatly favored by the intensive activity of military architects and engineers. The construction of new fortifications and the restoration of old ones, as well as the siege of a town, required accurate topographical site plans. Thus the science of surveying reached a higher degree of perfection through sheer practice and experiment as well as through the instruments invented by some of the same military architects.

Aided by their advanced scientific texts on geometry and driven by a need for secrecy, military architects tended to choose the representation in plan, that is, the horizontal section. This is the most abstract representation of a building or a city and consequently, the most difficult for an outsider to "read." The plan became, with the section, or profile, the preferred representational manner for military as well as civic architects, and became the shorthand for professional drawing, because of both its secretive and abstract qualities. The abstraction was the result of the use of accurate scales, also taught by military theorists, which placed each part of a building, fortification and city plan in precise relationship to one another, no longer privileging or neglecting parts of the object represented. The proportional plan provided information about the parts of the city which, through the use of scale, could be accurately measured off. The advantage of the abstracted plan was also that it allowed a

---

57 See Martha D. Pollak, "La storia delle città: testi, piante, palinsesto," *Quaderni Storici*, n.s. 67 (1988):223–56.

total, and unprejudiced view of the entire fortification and the city it enclosed, whereas the view always emphasized the objects in the foreground.

These improvements in surveying and draughting techniques, and the shift in mental attitudes implicit in the adoption of the horizontal and vertical sectional drawings, served to improve the art of cartography. The "histories" of military architecture published towards the end of the seventeenth century are, in effect, also cartographic albums. This is the case with the works of Nicolas de Fer, a cartographer and publisher, Bodenehr, an engraver and publisher, and Borgsdorff, also an engraver and a cartographer (cat. no. 21, 2, 3). Military architecture sped up the shift from the interpretive, but not always accurate—albeit visually powerful—view, to the abstract and totalitarian plan which unveiled all the secrets of a town as of a building, showing its inner and outer workings.[58] Although views of cities continued to be made and diffused to a public which often saw them as works of art, increasingly larger numbers of plans were provided, especially by de Fer, to satisfy scientific interest in the actual, accurate lay-out of foreign towns.

THE BASTIONED TRACE

Primers in geometry and graphic representation were merely the scientific instruments provided in the introductions of this group of treatises on military architecture. The main body of each work was invariably focused upon the principal interest of each military engineer and architect: the outline and detailing of the bastioned fortification trace. Although the principles and the fundamental parts of the bastioned trace were developed by the mid-sixteenth century, mostly in the work of Italian architects, some of the most fertile and inspired compositions—such as those by Francesco de' Marchi—were not published until the very end of the century. Thus the contribution of seventeenth-century theorists was not only to endow this new art with a scientific foundation achieved through codification and the application of systematic analysis, but to make it available to a larger public. In the section on bastions, which normally dominates in length and importance the structure of the military treatise, each author presents his fortification philosophy, discusses the internal problems of the profession's principal concern, and stakes his claim to original contributions in defensive strategy.

For instance, Pagan's greatest complaint about the fortification and defense of a city in the first half of the century was that, no matter how thoroughly reinforced and guarded, it could not resist a siege longer than six weeks.[59] His design efforts attempted to counter this universal condition, by widening the fortification belt and thus preventing the enemy's approach. Vauban based the design of his three "systems" of fortification upon Pagan's prescriptions. By the end of the seventeenth century he developed tables of calculations showing how long a place could resist according to its size and following his rules for provisioning it. His tables show conclusively that a fortified town in good

---

58 See Kim Veltmann, "Military Surveying and Topography, the Practical Dimension of Renaissance Perspective," *Revista da Universidade de Coimbra*, 27 (1979):263–79.

59 See Pagan, cat. no. 45.

condition could be reasonably expected to hold out under siege for 48 days.⁶⁰ Thus the net improvement 50 years after the publication of Pagan's treatise was of only six days of increased resistance. Furthermore, Vauban embraced unequivocally the paradoxical self-destruction at the core of fortification theory; he attempted to conceive of the perfect attack and to prepare the perfect defense. Although he, like most military theorists, did not often design the interior of his fortification enclosures—that is, the towns he fortified—his writings are especially concerned with containing a territory, as a recent analysis of his notion of the "pré-carré" convincingly demonstrates.⁶¹ The foundation of the post-cannon fortification trace was the bastion itself. Consisting of two flanks and two faces, it served essentially as a gun platform which provided flanking cannon fire that swept the adjacent curtain wall and the flank of the adjacent bastion. Thus the length of the curtain wall separating the bastions was determined by the shooting range of the available cannon and other firearms.⁶² Eventually the cannon could be turned around, aiming directly towards the countryside and shooting straight at the enemy rather than merely sweeping the walls of the fortification in order to prevent the enemy's approach. At that point the bastion became an aggressive element in the fortification which—after Pagan's innovations—began to proliferate and multiply in this double game of keeping the enemy further and further away from the walls and simultaneously using the cover of the walls to advance towards the enemy and break the siege.

Here too the treatise by Galileo provides a useful example of great clarity. The bastion is composed of gun platforms, scarp, parapet and "orecchione" defending the gun platforms. A dry or wet moat—or a combination moat—surrounds the bastions and the connecting walls, but also a counterscarp, covered way and glacis.⁶³ After the turn of the seventeenth century these basic layers of enclosure were reinforced with extensive outworks whose role was, in their turn, to defend the bastions, the gates and the curtain walls. Their forms evolved and multiplied throughout the century, and by the last third of the century Vauban used ravelins, half-moons, and counterguards, simple and double tenailles, hornworks and crownworks in Baroque profusion which transformed the defense in depth. By the end of the century this proliferation of fortification elements had had a great and general linguistic influence, and the names of the parts were fully ensconced in the imagination of a general public.⁶⁴

---

60 See Vauban, cat. no. 70, 71, 72.

61 Joan DeJean, op. cit., ch. 1.

62 For the history of the bastion's development see, among others, John R. Hale, "The Early Development of the Bastion: an Italian Chronology, c. 1450-c. 1534," in *Europe in the Late Middle Ages*, ed. Hale et al. (Evanston, 1965), 466–94; Quentin Hughes, *Military Architecture* (London, 1974), ch. 3; Ian V. Hogg, *Fortress, a History of Military Defence* (London, 1975), ch. 3.

63 For a glossary of fortification terms see Christopher Duffy, *Fire and Stone; The Science of Fortress Warfare 1660–1860* (London, 1975), 183–86; among these treatises see Dögen, cat. no. 15, 293–95; Cellarius, cat. no. 9, prefatory matter; Freitag, cat. no. 24, 25, book 1; Capra, cat. no. 6, part 2.

64 Indeed "by the middle of the seventeenth century, the art of fortifications was no longer a mystery to the general public" (quoted from *Blomfield* in Jean DeJean, op. cit., 28), while these

The best known example of a military buff is Tristram Shandy's uncle Toby, a wounded survivor of the battle of Namur. He begins his studies of siege and fortification in order to exorcise his fears and heal his wound, but they soon

---

verses published by Abel Boyer in *The Draughts of the Most Remarkable Fortified Towns of Europe* (London, 1701) convey vividly the interest and richness of the established language of military architecture. Its expression in French shows the preeminent influence of Vauban's principles and successes, 9–10:

Dans les siecles premiers rien ne *flanquoit* les *Forts*:
On ne pouvoit braver l'insulte du dehors;
On connut ce defaut, on fit des *tours quarrés*,
D'un petite intervalle entr'elles separées;
Et la suite du tems, qui deffile les yeux,
Arrondit ces quarrez pour *flanquer* beaucoup mieux.
Mais depuis *Berthold* eut inventé la *poudre*,
A chercher d'autres *flancs*, il falut se resoudre.
On fit des *bastions*, autrement *boulevards*,
Qui font les vrais soutiens des *murs* et des *ramparts*.
Ils ont *angle flanqué*, *face*, *angle de l'epaule*,
Flanc, *ligne capitale*, et *gorge* à double rolle;
La *courtine* placée entre deux bastions,
A quelquefois deux *flancs*, que l'on nomme *seconds*.
La *ligne de defense* en tel cas est *fichante*;
Cette ligne autrement n'est jamais que *razante*.
L'*angle diminüé*, l'*angle* appellé *flanquant*,
L'*angle de contrescarpe*, et *saillant*, et *rentrant*;
L'*exterior côté*, le *double diametre*,
Sont des *termes de l'art* qu'il ne faut pas omettre.
L'*angle du centre* encore doit être seu par coeur,
Et vous devez sur tout supputer sa *valeur*.
Pour la trouvez voyez la somme resultant
Du nombre des côtez, coupant trois cens soixante,
Et de cent quatre vingts l'*angle du centre* ôté
Donnera la valeur de l'*angle du côté*.
Le *flanc* du grand *Vauban* en cercle se figure,
Il touche l'*orillon*, il touche la brisure;
Autrefois on faisoit, peu raisonnablement,
Un orillon *quaré* qu'on nomme *epaulement*.
Quant à ces trois *dehors*, *ravelin*, *demilune*,
Contregarde, ils ont tous quelque chose commune,
Qui fait qu'ils peuvent être aisément comparez
Avec des *bastions* du *rampart* separez.
Pour la *tenaille simple*, et la *double tenaille*,
L'esprit les voit bien-tôt, pour peu qu'il travaille.
La *simple* qu'il connoit en faisant moins d'effort,
Avec l'*ouvrage à corne* a toûjours grand rapport.
La *double* qui ressemble à l'*ouvrage-à-couronne*,
Ne passe point par tout pour defense fort bonne.
Le *corridor*, qu'on fait quelquefois à *redans*,
Est le premier *dehors* ou vont les assiegeans.
L'*esplanade* ou *glacis* le suit, il environne,
Tous les autres *dehors* ainsi qu'une couronne.
Voilà quels sont les noms des principaux *dehors*.
Contre qui l'assiegeant fait ses premiers efforts.
Le *dessein* ou le *plan* se nomme *ichnographie*;
La hauteur ou *profil* s'appelle *orthographie*.
Sachez encor le mot de *parapet*, *cordon*,
Poterne, *fausse-braye*, *embrasure*, *merlon*,
Casematte, *talud*, *plateforme*, *banquette*,
Pas de souris, *gazon*, *terre-plain* et *cunette*;
Sachez que le hauteur qu'on nomme *cavalier*
Est un second *rampart* assis sur le premier.
La *cascane* est un puit creusé contre la *mine*;
Les *portes* ont *bacule*, *orgues*, *pont*, *sarrasine*.
L'*escarpe* et *contrescarpe* au *fossé* se font voir,
Et ce sont deux *taluds* par où l'on y peut cheoir;
Chaussetrape pointue, épaisse *barricade*,
Fraise, *cheval de frise*, et haute *palissade*
Arrêtent l'ennemi dans son premier effort,
Embarassent ses pas, et lui ferment l'abord.
Dans un *siege* on se sert de *fortins*, *gallerie*,
Boyaux, *hute*, *barraque*, et *chars d'artillerie*;
On y remarque encor *circonvallations*,
Tranchée et les detours, *chandeliers*, *gabions*;
Grenade, *pots à feu*, *mantelets*, *sacs à terre*,
Mais ce qui montre mieux la vigueur de la guerre,
On y voit le *petard*, joint à son *madrier*,
Le *boule* au *canon*, et la *bombe* au *mortier*.
L'effroyable *carcasse*, invention recente

become a hobby-horse. Not only does he become an expert of military strategy, but his enthusiasm and devotion are so contagious that they contaminate the entire Shandy household and especially corporal Trim, his valet, who gamely offers to build for uncle Toby a model of the Namur siege in the garden.[65]

The military authors represented here attempted to introduce refinements and improvements in the design of the bastioned trace, from its initial formulation at the beginning of the sixteenth century through the constructions of Vauban and his contemporaries. The widening of the fortification belt was the most visible result of this elaboration. The reasons for layering the fortification defenses were based on the need to distance the enemy as far as possible from the walls of the city and to create an image of the fortification so awesome that potential enemies would be discouraged from siege by the sheer sight of it. It is the latter intention that is most evident in the treatises on military architecture, where an aesthetic of fear is developed through increasingly astounding drawings whose intention is to represent fortification that would inspire awe and strike terror in the onlooker.

The perspectives by Speckle are certainly very successful in inspiring terror, not only because his fortification is powerful—endowed with double-platformed bastions, with scarp and cavaliers and a highly articulated covered way and glacis—but also because it was effectively drawn in orthographic and scenographic representation. These scientifically conceived and powerfully illustrated fortresses are in strong contrast with his romantic castles raised upon unlikely and unreachable mountain tops, probably inspired by Dürer's landscapes but also partially realized in the hills south of Strasbourg (cat. no 55). Zanchi illustrates well the staggered curtain wall which allows for additional gun emplacements and for greater distance between the projecting bastions (cat. no. 73). This idea is also proposed by Girolamo Cataneo and Errard, while Maggi's curtain walls are angled almost

---

> Remplit dans les citez les peuples d'epouvante;
> La foudre ne fait point de ravages plus grands,
> Que ces traits enflammez qu'on veu les derniers tems.
> Il vous faut bien connoître exactement ces choses,
> Leur matiere, leur nom, leurs effects et leurs causes;
> Car, entre tous les arts, *fortification*
> Demande à vôtre esprit grande application.
> Il faut étudier d'abord la *reguliere*,
> Qui fournit a voir l'autre une grande lumiere;
> Pour l'avoir dans l'esprit fort present en tout tems,
> Retenez avec soin les terms *grecs* suivans:
> *Penta* c'est cinq; *ex*, six; *epta*, sept signifie;
> *Octo*, huit; (le *Latin* au *Grec* ici s'allie.)
> *Ennea*, donne neuf; *deca*, veut dire, dix;
> Onze, c'est *endeca*, *dodeca*, deux fois six:
> A chacun de ces mots joignez celui de *gone*,
> Et vous aurez le nom de chaque *polygone*.
> Pour les fortifier des sentiments divers,
> On ne peut qu'en passant vous parler dans ces vers
> *Errard*, que maintenant en tous lieux on surpasse,
> Joignoit par *angle droit* le *flanc* avec *la face*;
> D'autres pour decouvrir beaucoup mieux l'aissaillant,
> Joignent par *angle droit* la *courtine* et le *flanc*.
> La *ligne de defense* est *perpendiculaire*
> Sur le *flanc* de *Pagan*, autheur que l'on revere:
> Et dans ces derniers jours le célèbre VAUBAN
> A perfectionné le *dessein de Pagan*.

[65] Laurence Sterne, *The Life and Opinions of Tristram Shandy, Gentleman* (Oxford [1760], 1983), Volume II, 65–122.

to become projections of the bastions themselves (cat. no. 7, 19–20, 35). Lorini, Alghisi, Tetti and Sardi flank their bastion at its neck—where it is connected to the curtain walls—with twin cavaliers (taller towers raised above the bastions) whose cannon are intended to sweep the countryside directly (cat. no. 32, 1, 66, 52). In Tetti's design the twin cavaliers are echoed by crownwork outworks which protect the bastion from the outside. Groote's fortification relies entirely on various pieces that comprise the outworks; Tensini's bastions project far beyond the curtain wall extended with a wealth of crownworks, while Goldmann's fortress is layered as much in height as in width, providing the defending garrison several platforms from which to repel the enemy spread out before it (cat. no. 28, 65, 27).

This layering in depth achieves new dimensions in the proposals of Pagan and Marolois, while Dilich's graphic technique makes up in ferociousness where the actual fortification might fail (cat. no. 45, 40, 14). Cellarius' fearsome fortification though clearly inspired by Speckle's compositions and representational manner, expands greatly upon his chosen model (cat. no. 9). Coehoorn's design leaves his predecessors' conceptions far behind him: inspired by their studies of polygonal ideal fortifications he proliferates bastions, moats, ravelins, half-moons, crownworks and covered ways with such prodigality that the result is a delirium of fortification run amok, an inflationary approach in which he is not, however, alone (cat. no. 10).

The reality and actuality of military architecture are illustrated in the treatises published towards the end of the seventeenth century by de Fer, Ozanam, Mallet, and Borgdorff. These are a different kind of publication: although their purpose is still to educate in military architecture their teaching is done in a historical manner, by providing examples of the great realized fortification projects rather than attempting to document a personal innovation and to persuade the community of military theorists of its originality. The illustrations that form the bulk of these works show the extent to which the ideas contained in the earlier treatises had actually been realized in bricks and stone. The plans and views of places as culturally and geographically varied as Palmanova, Antwerp, Berlin and Turin, Lille, Coeworden and Mannheim (cat. no. 15, 24, 2, 21)—among others in this catalogue—show that the widely accepted bastioned fortification surrounded most western European towns in the seventeenth century. They also document the proliferation of fortification outworks which isolated the city turning it into an island. These illustrations show the city freestanding within a wide belt of open land, cleared of buildings and trees in order to provide unobstructed sightlines for the defenders and mined during siege in order to destroy careless attackers. This menacing isolation of the city inside its mined belt and bastioned trace was effectively represented in the early histories of fortification, which thus publicized the terrifying character of seventeenth-century permanent fortification of cities.

These "historical" fortification treatises became invaluable for the education of the military architect and indispensable for the training of military strategists, and by representing an awesome military machine could act as a deterrent force. Their authors concluded

xxxiii

the two-century-long effort that rendered military power scarily beautiful. Furthermore, these books provided the aesthetic lens through which fortifications were to be viewed, they thus ennobled them with formal qualities that were eventually adopted also in civic architecture.[66] They promoted the work of military architects by codifying the disparate fragments of military defense, and they raised siege warfare to the level of science.

Attack and Defense

Siege was the quintessential means of warfare in the seventeenth century, especially in the numerous wars waged by Louis XIV, Vauban's sole patron. The king is said to have hated the surprise and chance of the battle in the open field where he could expose himself to failure; preferring a complete and accurate organization he favored sieges instead, because there everything was organized in advance.[67] Vauban was considered a theoretical, systematic and machinating genius, and therefore much more threatening than his most talented contemporary opponent, the Dutch strategist Coehoorn. His tables of calculations gave the impression of strategic unassailability; since he calculated not only the dimensions of every element of the fortification, but also the length of time it would take the enemy to gain individual layers of the fortification, every stage of the siege could be predicted in advance. Vauban reduced the defense and attack of fortresses to double-entry bookkeeping, where the two columns balance each other precisely. The accountability of the smallest part of the defense, fortification and provisioning in Vauban reflect the earlier attempts by military theorists to set up a machine which can be expected to operate by itself, but which resulted—both in Vauban and his predecessors—in an obsession with the smallest detail.

It was Vauban who made attack and defense indistinguishable from one another, and inseparable.[68] But this quest for invulnerability required constant updating since the two aspects of siege warfare were never at parity—the advantage was almost throughout with attackers because siege strategy was favored by innovations in firearms, and it was a commonplace (certainly after Pagan publicized it) that all fortresses could be taken in time—and thus the design of the perfect defensive fortification was constantly reformulated. The indivisibility of attack and defense produced the proliferation of defensive fortification illustrated in the theoretical proposals as well as the realized enclosures. The outworks were endlessly multiplied, bristling with layer after layer of half-moons, ravelins, tenailles and hornworks; increasing their terrifying

---

66 For the influence of military architecture upon civil architecture see Christian Otto, "L'Architecte et l'ingénieur: Neumann et la Residenz de Werneck," *VRBI* 11 (1989):71–84, and Wilhelm Waetzoldt, *Dürer and His Times* (London, 1950), 224: "Dürer's study of the theory of military architecture is a presage of a new development which was to appear in the period of the German Baroque—military engineering as a basis of monumental architecture, with military engineers and artillery officers (Speckle, Dientzenhofer, Neumann) as masters of sacred and profane architecture."

67 Joan DeJean, op. cit., 31.

68 ibid., 53: "To pit the perfect attack against the perfect defense is to put into operation a military machine that will function until not a single combatant remains."

aspect by seeming to come alive, despite their foundation in geometric, rather than so-called natural forms, they engendered the horrifying impression that fortification is an endless project, rendering it even more dangerous. Although this proliferation could occasionally seem out of control, it imposed a geometrization of space through suggested patterns which, paralleled in French seventeenth-century gardening methods, proposed an ordering system that could have also tranquilizing effects.[69]

THE CITADEL

The ultimate protection provided by the citadel—a fortress within a fortress—was the encouragement of last resort offered by military architecture. Although considered morally reprehensible already in the sixteenth century,[70] many citadels were built especially in Italy, northern France and Germany. The citadel had the double function of defending the city from outside attack—both Francesco di Giorgio Martini and Gabriele Busca considered it the head of the city which could be taken only after the city itself had been defeated[71]—and of defending itself from the inhabitants of the city. The citadel thus illustrates the social hierarchy and conflicts inherent in early modern European governments. The citadel seems to disappear after the mid-seventeenth century when in Coehoorn and Vauban's conception each fortified town becomes a citadel.

Placed at the center of the city in the earlier conceptions of Leonardo,[72] it is dislocated to the edge in realized sixteenth-century versions, thus illustrating its equivocal relationship to the city itself (cat. no. 48, 21). Invariably smaller than the city, the citadel was a regular geometrical form, often a pentagon, which has remained a form associated with national military enterprises to this day.

Composed of a single geometrically-pure form, containing a city's best ammunition and devoted entirely to its military presence, the citadel is the paradigmatic representation of the fortress modelled upon the ideal city—Euclidian form and central planning, with total social control represented in its architectural composition. Together the bastioned trace, the outworks and the citadel were calculated and designed to resist the most belligerent, innovative and patient siege. Both sides of the war, siege and defense, were designed by the same military architects and engineers, who throughout their careers took turns in being alternately attackers and defenders. This paradoxical, bilateral mental state is the most outstanding characteristic of the seventeenth-century military treatise, and it creates an unsolvable tension that is tangible both in the texts and their illustrations.

* * *

Fortification treatises, then, formed a literary type defined by shared interests and purpose, conventions of technical vocabulary, and structure of arguments. They represented a vivid interest shared by heads of states and strategists who engaged in war—a widely shared experience and condition during the

---

69 ibid., 64–65, for a discussion of Louis XIV's *Manière de montrer les jardins de Versailles*—his itinerary for the right way to visit the royal gardens—and the tranquility induced by Le Nôtre's "militaristic tabulation of nature."

70 See above n. 17.

71 Busca, cat. no. 5, preface.

72 See above, Marani, n. 12 and Heydenreich, n. 2.

xxxv

two centuries examined here. Besides reflecting certain key concerns of the Renaissance and the Baroque, these texts contain an internal discourse focused on war and on each other. The treatises were crucial in the education of architects, engineers, artillery officers, generals and monarchs as well as eventually influencing the education of an important intellectual elite. The forms of military architecture and the language of its strategic theory eventually altered mental perceptions of the city, literary discourse, social relations and the surroundings of the urban dweller. Evidence of the widespread acquaintance with military and fortification terminology—predominantly concerned with attack and defense—can be found in many other kinds of literary production, among them utopian philosophy, pedagogies and libertine manuals (*Les Liaisons Dangereuses*, the best known among the latter texts, was written by an artillery officer, Choderlos de Laclos), while the perfected Vaubanian fortress, the codified supreme achievement and conclusion of previous innovations in military architecture, came to stand as a symbol of French cultural achievement. The defensive imagination applied to securing the impregnability of the fortress was translated into a mental habit of accounting and enclosing which eventually pervaded the approach to every human endeavor—love, literature, public relations. Thus the treatises on military architecture had a theoretical semantic impact on other endeavors as well as being the foundation for the practice of fortification construction, attack and defense.

CATALOGUE

1 **Alghisi, Galasso,** c. 1523–1573.
Delle fortificationi libri tre.
[Venice] 1570.

26.4 × 40 cm. Brown calf, goldstamped title on morocco panel, Kimbolton Castle shelf label inside upper cover.
lxii + 406 pages. Table of contents, subject index, preface, 39 engravings in Book 2, 192 × 202 to 524 × 400 mm.
First edition. The dedication to the Holy Roman Emperor Maximilian II is dated 28 November 1570.
REF: HCL, DBI, Miscellanea 14
COPIES: MH, BM, BN, Bodleian, NN, MiU, ViU

GALASSO (OR GALEAZZO) ALGHISI DA Carpi was a military and civil architect. Between 1544 and 1558 he was employed—together with other architects such as Antonio da Sangallo—in the construction of fortifications commissioned by Paul III in Rome. His experience in civic architecture included the construction of the sanctuary in Loreto (1549), the centrally-planned Santa Maria delle Vergini (1550) and a design for the municipal tower (1558), both in Macerata. After 1558 he passed into the service of Ercole II and Alfonso II d'Este in Ferrara. Alghisi describes himself as the architect of the duke of Ferrara, and his design for a palace for Alfonso II is extant in an engraving by Domenico Tibaldi.

His treatise, divided into three books, is focused entirely on fortresses. In book one, subdivided into fifteen chapters, Alghisi discusses the state of the art, criticizing—inter alia—the work of Maggi and Castriotto and the little resistance that contemporary fortifications were able to offer to artillery. Book two, of 98 chapters, deals with the composition and form of fortresses in geometrical order and size, from the pentagon to the 21-sided fortress, while book three—19 chapters—is about the site, the materials and instruments needed for the construction of the fortification.

In his preface Alghisi addresses the two principal problems debated in professional military architecture in the sixteenth century: should the city be fortified, and why the *ingegno* of the architect is a key element in the planning of defensive and offensive strategy. Alghisi urges that the design of fortresses be made first on paper since it is easier to see and solve errors in design. He insists that practice in war does not provide enough training for the design of fortresses, which requires knowledge of geometry and arithmetic as well as perspective, and a reasoned method of fortification. Alghisi maintains that fortresses on a flat site in the plain are better than those on hill sites (contrary to Castriotto). The angle of his suggested curtain walls results in a deeply indented perimeter, thus a large fraction of the enclosed territory is lost to fortification. Alghisi does not examine the construction of the bastion and wall, nor does he propose a layout for the interior of the fortification enclosure.

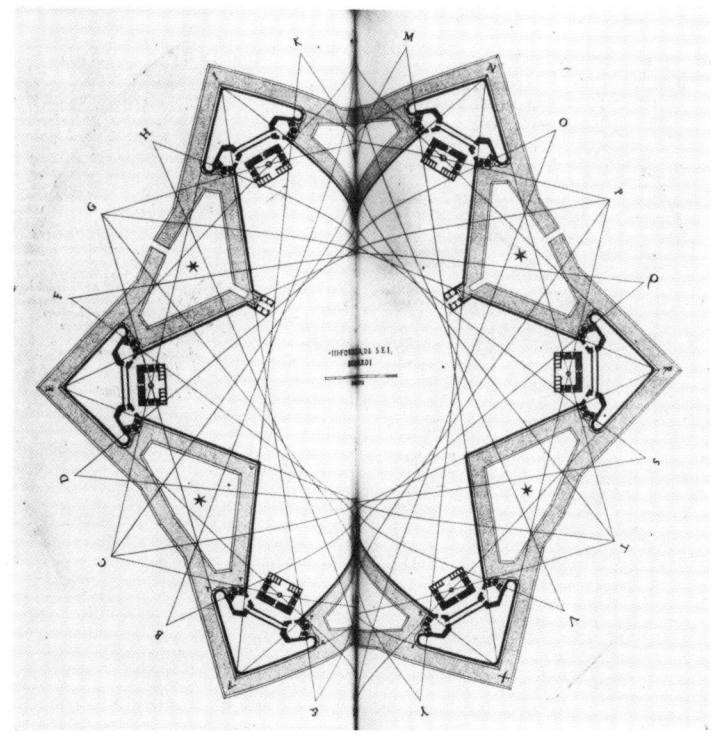

2 **Bodenehr, Daniel,** 1664–1758.
Force d'Europe, oder die merckwürdigst und fürnehmste, meistentheils auch ihrer Fortification wegen berühmteste Staette, Vestungen . . . in Europa.
Augsburg, Daniel Bodenehr, n.d. [but after 1708].

32.3 × 20 cm. Half brown calf, speckled black and tan boards, goldstamped title on panel, arms of Johann Georg Baron of Pichelstorff and Altenburg at center of covers.
ii + 200 leaves + i. Engraved title-page, 200 engraved plans of 150 European cities and fortresses, 323 × 200 to 530 × 200 mm, list of illustrations.

First Edition.

REF: ADB
COPIES: NN

THE BODENEHRS WERE A FAMILY OF ENgravers active in Augsburg and Dresden from the mid-seventeenth to mid-eighteenth century. Each of the 200 numbered plates is signed by Bodenehr, whose enterprise is similar to that of Nicolas de Fer, as is the title of his publication. Many plates are accompanied by a historical description of the city and a legend of principal buildings.

BERLIN.
DOROTHEEN STATT.
FRIDRICHS WERDER.

MÜNCHEN
Die undere blaich
Wasser thurm
Kalch Offen
Gries
Aufm Gries
H. Geist Anger
Unsers Herrn Thor
Hertzogen Statt Thor
Anger Thor

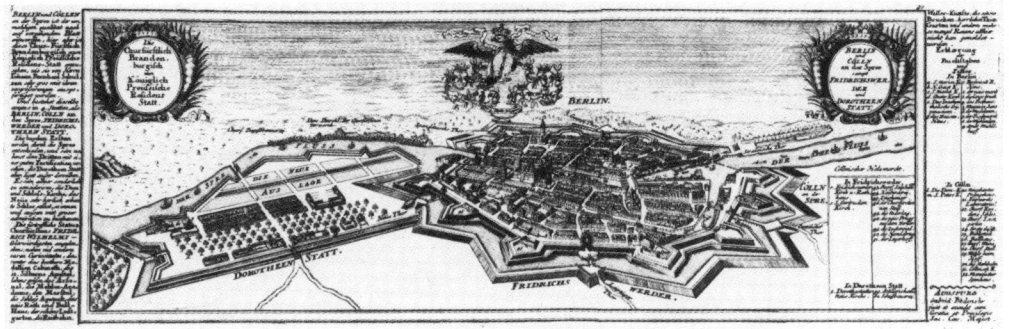

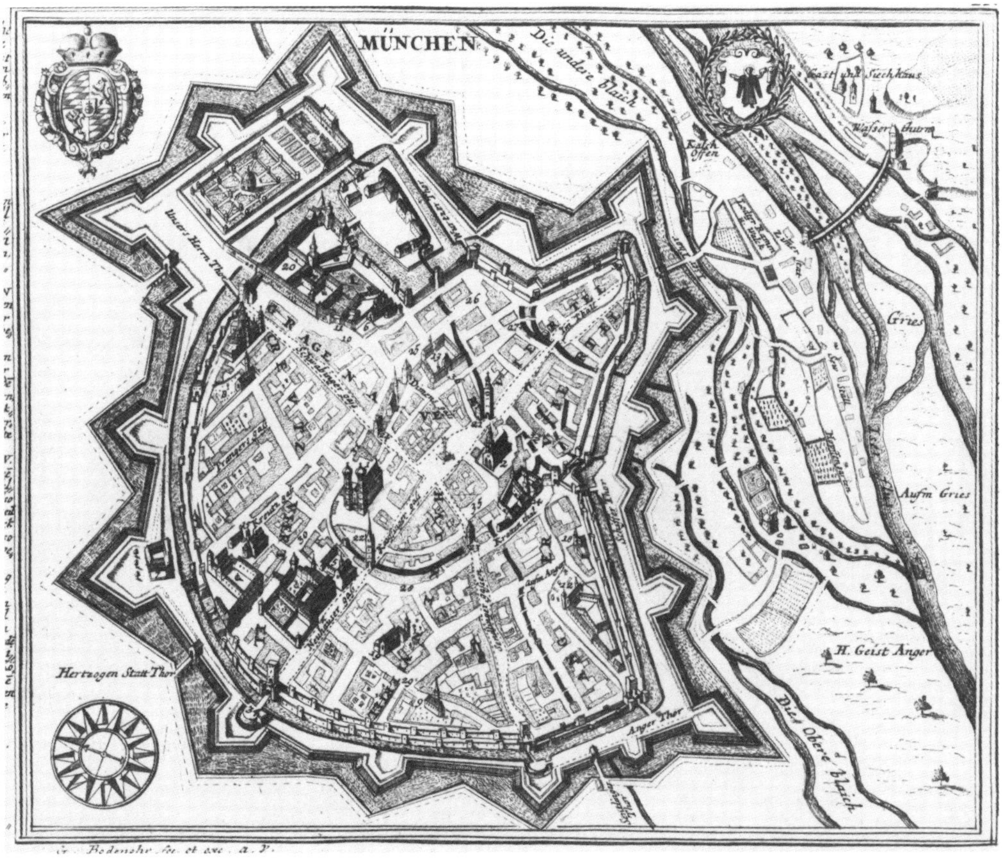

3 **Borgsdorff, Ernst Friderich, Baron von.**
Academia Fortificatoria, oder klare Instruction wie man die Stätt auss approbirten Kriegs-Gründen bevestigen/beschützen und bewältigen solle.
Vienna, Johann Van Ghelen, 1694.

7.4 × 12.8 cm. Stained limp vellum, blue edges.
xx + 250 pages. Engraved title-page, introduction, table of contents, 10 engraved illustrations folded in, 180 × 128 to 182 × 128 mm.
First Edition.
REF: Marini

DIVIDED INTO TWO PARTS, THIS TREAtise deals with the history of fortification, the design of regular fortresses, siege, the irregular fortification enclosure, assault, and field fortification. The illustrations are labelled in Italian and in German. Borgsdorff, an engineer in the service of the Holy Roman Emperor, is also the author of the album *Neu-triumphirende Fortification* (Vienna 1703) in the Newberry collection.

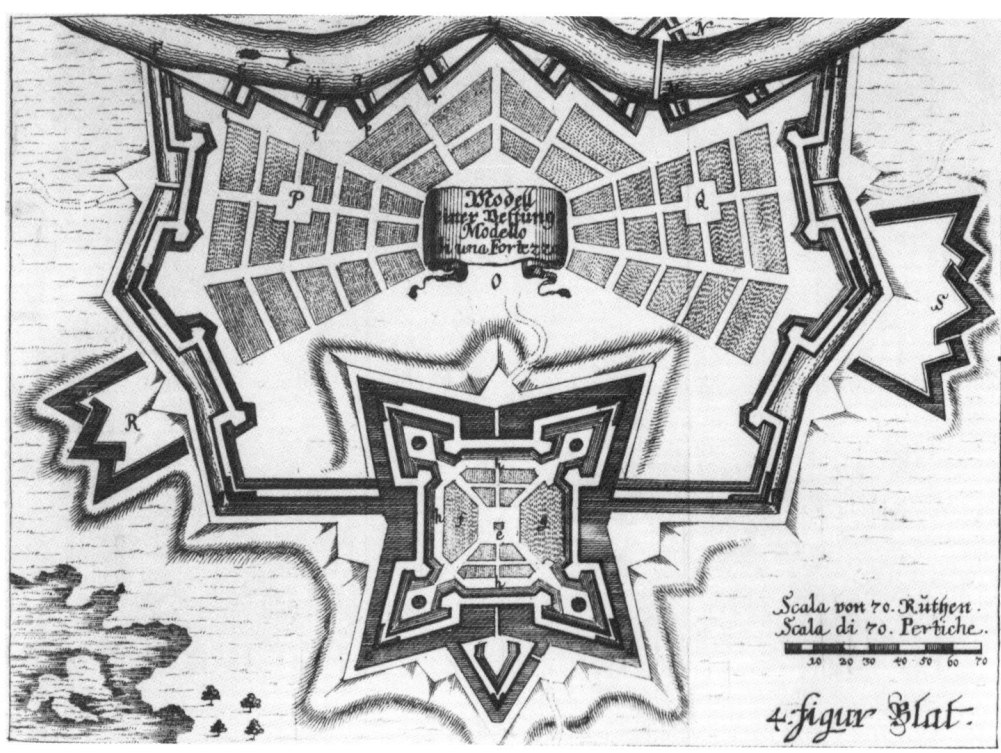

4 **Bourdin, Pierre,** 1595–1653.
L'Architecture militaire, ou l'art de fortifier les places regulieres et irregulieres.
Paris, Guillaume Benard, 1655.

10.6 × 16.6 cm. Contemporary limp vellum, with calculations written on upper cover. 196 + 35 pages. One engraving tipped-in before title-page, 73 full-page woodcuts in text, 71 × 122 mm, appendix titled "La Géometrie militaire" with 24 half-page woodcuts, 75 × 40 mm.

First Edition. Published with royal license and permission from Louis Cellot, "provincial" of the Company of Jesus in France, dated 9 February 1655; dedicated to the French nobility.

REF: DBF
COPIES: MiU

A MATHEMATICIAN AND A JESUIT, Bourdin taught at La Flèche and Paris, where he was a member of the Company of Jesus headquartered in the Rue St. Jacques. He published numerous school texts, including a treatise on the sun. Known for his attacks aimed at Descartes, whose *Optiques* he criticized, accusing the author of scepticism (Descartes replied at length and complained to Bourdin's superior about these attacks).

Bourdin's treatise was published posthumously, after it had been reviewed by three members of the confraternity. He dedicates most attention to the terminology of fortification, to its geometrical formulation, and to a comparison between the French, Italian and Dutch methods of fortification. In his discussion of the "ordres de la fortification" (p. 66) Bourdin makes an analogy between the orders of architecture—named after their Greek and Roman regions of origin—and the "orders" of fortification named after the three main national contributions, French, Italian, and Dutch.

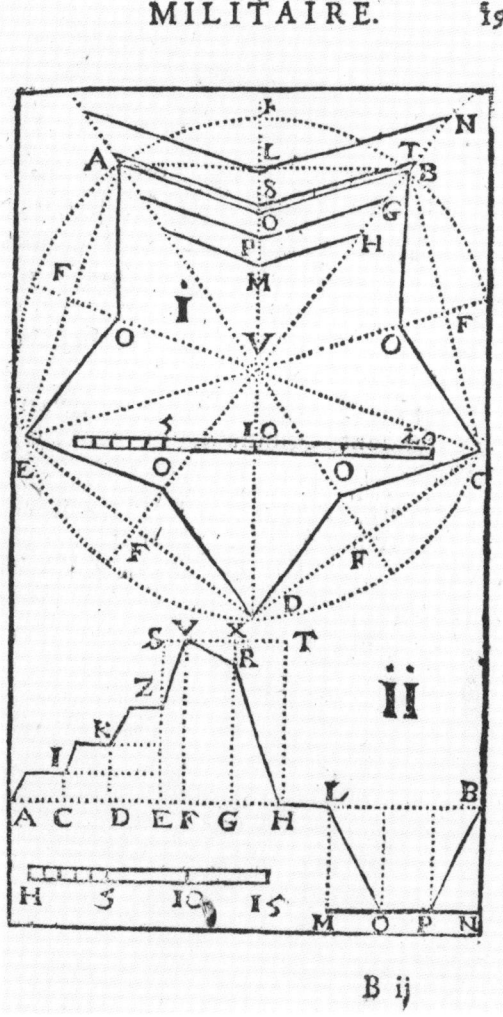

5 **Busca, Gabriello,** c. 1540–1605.
Della espugnatione et difesa delle fortezze libri due.
Turin, heir of Nicolò Bevilacqua, 1585.

18.3 × 23.6 cm. Contemporary worn and distressed vellum, two pairs of ties, one removed. Numerous manuscript marginalia in Italian.
viii + 256 pages + iv. Table of contents, errata page, ten double-page woodcuts of battle scenes and fortifications, 348 × 236 mm. They are bound as folded leaves in this copy and included in the pagination.
First Edition. The dedication to Carlo Emanuele, Duke of Savoy, is dated 1 January 1581 from the suburb of Brescia.
REF: HCL, DBI
COPIES: MH (1585, 2d edition, Turin 1598), BM (1585, 1598), DLC, NN (1585, 1598), NNC (1585, 1598), CLSU (1598), MiU (1598), MnU (1598), RPB (1598), PPULC (1598)

Son of a founder of artillery employed by the state of Milan, (Gabrio) Busca was employed as military architect and engineer at the Savoy court from 1570 to 1594. Responsible for all military construction in Savoy, he completed the fortification of Bourg-en-Bresse begun by Francesco Paciotto. In 1592 he built the fortress of San Francesco on the border with Dauphiné, restored Montmélian, and built the forts at Demonte and Susa. In 1593 he took part in the successful siege of Exilles. After 1594 he was in Spanish service in Milan, campaigning against Henry IV in Burgundy until 1598; named "architect of the state of Milan" in 1599, he inspected the artillery of Pavia, Valenza, Novara and Alessandria. Commissioned by the governor of Milan Luis de Velasco, Busca copied some of Francesco di Giorgio Martini's drawings of ancient war machines (preserved at the ducal library in Turin) requested by Justus Lipsius for his *Poliorceticon*. Busca probably influenced the military education of Carlo Emanuele I, duke of Savoy, who studied the manuscript of this treatise from 1578 on. Beside his actual involvement with war actions, Busca experimented with *regole belliche*; in 1584 he tested Tartaglia's, Mora's and Cardano's ballistic prescriptions by shooting against the walls of Turin.

Two short MS works by Busca are preserved at the Biblioteca Trivulziana in Milan: "Discorso di fortificazione all'Ill[ustrissi]mo signor marchese Carlo Filiberto d'Este" and "Discorso sopra le misure delle cortine, fianchi, e spalle de' baluardi d'una fortezza reale." His published literary contributions include a treatise on artillery, *Instruttione de' bombardieri* (Carmagnola 1584, Turin 1598), and one on the design of fortifications, *L'architettura militare* (Milan 1601, 1619), announced in the first edition of the *Espugnatione*.

Busca's treatise is premised on the postulate that all fortresses, no matter how strong and well-sited, can be taken. His maxims include the beliefs that the teaching of offensive attack will also teach the necessary means of defense, that, although the citadel is like the head, of the city which forms the body (deriving this notion from earlier writers like Francesco di Giorgio), the latter should be taken before the former can be besieged successfully, and that the form of the fortress is more important than the construction material and the thickness of the wall. He does nonetheless broach the problem of materials, bringing up a subject still debated in the last quarter of the century: whether fortifications should be of earth or of masonry. Busca reiterates the problems of earthwork fortification (a longevity of only six or seven years) and seems content with the solution—proposed earlier in the century—to clad the earth wall with brick. He counsels a thin cladding, however, and a wall that leans inward from the moat so that if the brick falls into the moat, presumably under the impact of cannon battery, it will not suffice to make a bridge for the enemy.

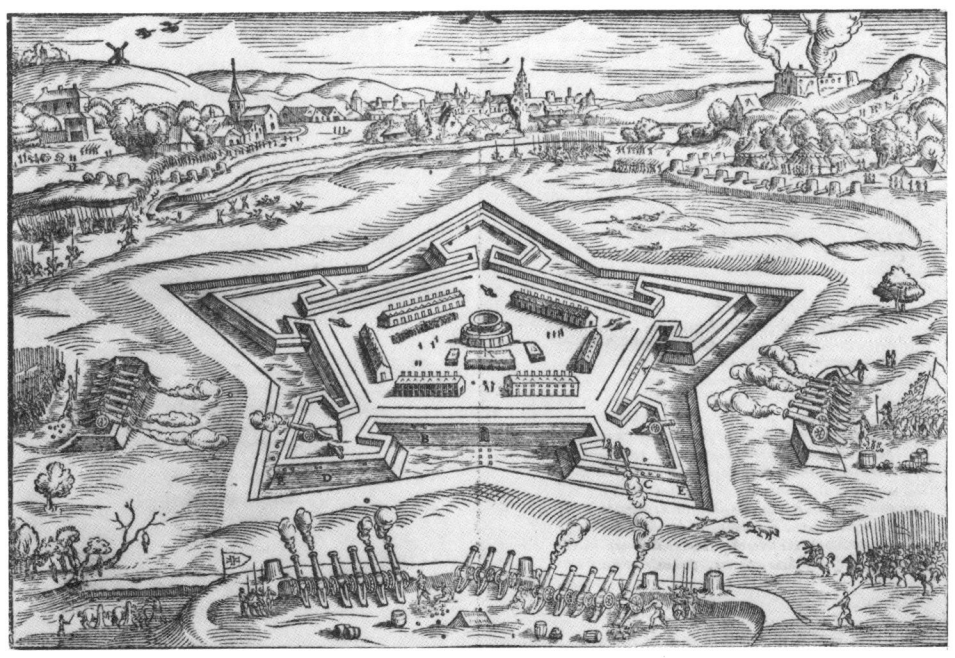

6 **Capra, Alessandro,** c. 1609–c. 1683.
La nuova architettura civile e militare.
Cremona, Pietro Ricchini, 1717.

17.5 × 22.2 cm. New binding, half leather, brown paper boards.
356 + 184 pages. Woodcut portrait of Capra at 70, 7 woodcut part title-pages, two prefaces, table of contents for book two, numerous woodcut illustrations in part one, subject index; 92 woodcut illustrations in part two, 84 × 81 to 384 × 184 mm.

Second Edition. Dedication to Doctor Francesco Arisi of Cremona is dated 25 January 1717.

REF: Wiebenson, DBI, Dezzi-Bardeschi
COPIES: BN (1683, *La nuova architettura militare d'antica rinovata*), CtY, N, NjP, NN, IU

An able stage and fountain designer as well as architect and engineer, Capra came from Cremona and studied with the military architect Jacopo Erba, spending time in Milan between 1628 and 1630, and working on the cathedral of Pontremoli (1630). Between 1647 and 1648 he worked on the defense of Cremona during its siege. His earlier publications include *Geometria famigliare, ed istruzione pratica* (Cremona 1671), *Nuova architettura dell'agrimensura di terre ed acque* (Cremona 1672), and *Le due prime parti della geometria famigliare* (Cremona 1673).

The Newberry copy brings together his two major publications, *La nuova architettura famigliare,* first published in 1678 in Bologna, and *La nuova architettura militare d'antica rinovata,* first published in Bologna in 1683. *Architettura famigliare* is divided into five parts corresponding to the five orders of architecture, but this is a purely symbolic arrangement, since the actual content of the parts is focused upon estate planning and planting, surveying, estimating construction costs, the principles of geometry, and the design of various machines. *Architettura militare* is divided into three parts. The first deals with the principles of geometry, which are then illustrated in Dutch fortifications. Part two deals with the Italian style of fortification, while part three deals with mechanics. The woodcuts are very coarse and undetailed, but the machines they represent were considered very innovative and established a high reputation for the author.

Delle Fortezze all'Italiana.

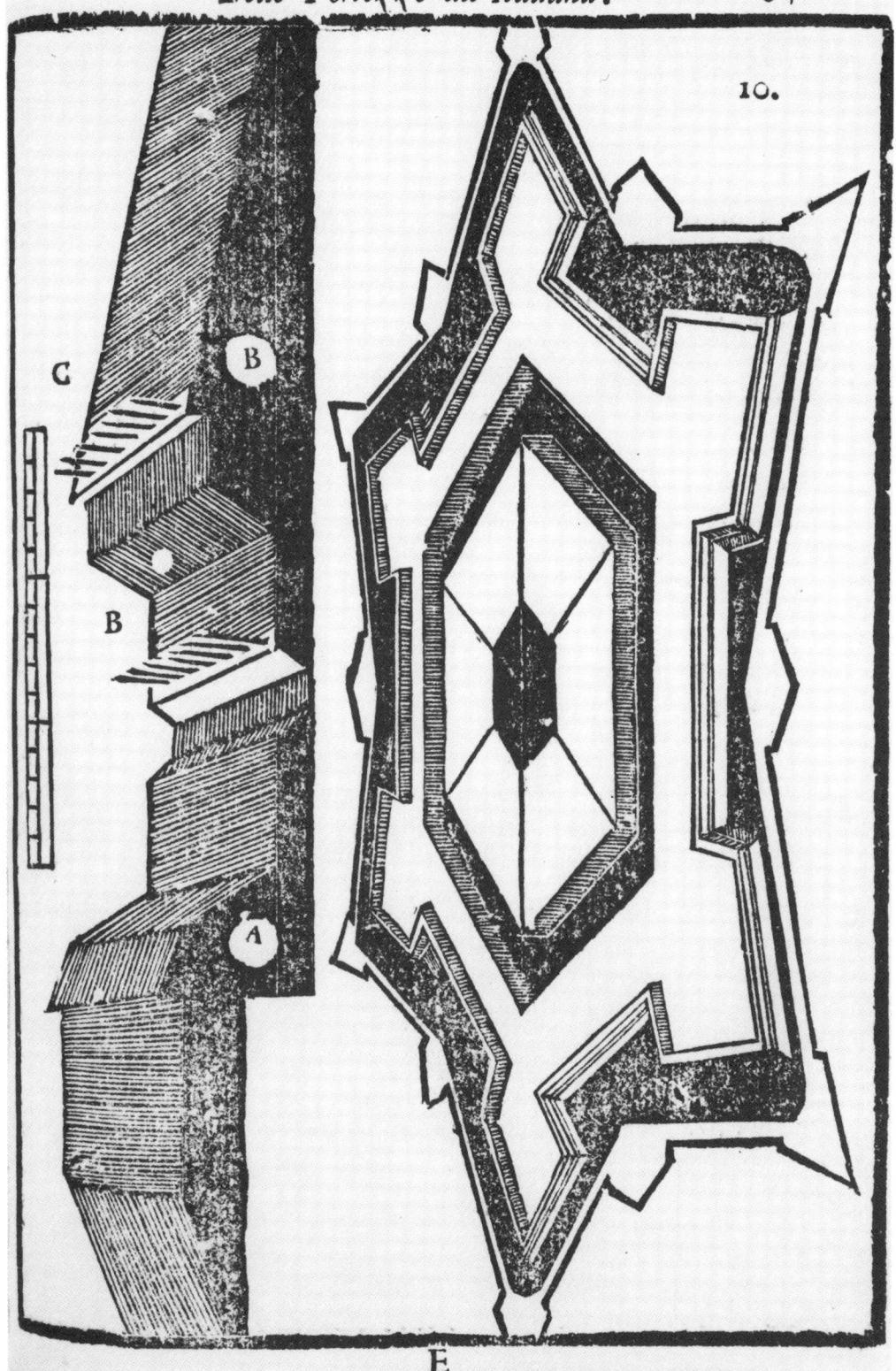

7 **Cataneo, Girolamo,** active c. 1540–1584.
Dell'arte militare libri cinque, ne' quali si tratta il modo di fortificare, offeendere (sic), et diffendere una fortezza.
Brescia, Thomaso Bozzola, 1584.

15.2 × 20.5 cm. Contemporary vellum, worn and shrunk, two pairs of ties removed.
iv + 80 + 35 + 30 + 29 + 39 leaves + i. Engraved title-page and 4 part title-pages, 95 woodcut illustrations in the first three books, mostly geometrical diagrams, bastion studies and siege representations (20 double-page illustrations in book one, 6 illustrations folded in), 65 × 36 to 522 × 205 mm; book one: table of contents, preface and introduction; book two: preface, book three: table of contents, preface; book four: preface, tables of calculations; book five: table of contents, introduction; this copy is bound with Girolamo Cataneo's *Dell'arte del misurare libri due*, Brescia, Thomaso Bozzola, 1584.

Fourth Edition.

REF: HCL, Cockle, DBI, Miscellanea 12, Fortifications
COPIES: MH (1564, 1567), BM (1571, 1584 [but Pietro Maria Marchetti publisher], two French editions, 1574, 1593), Folger (1608), MiU, DLC, IU

CATANEO OF NOVARA WAS IN THE SERVICE of the counts d'Arco in 1542 in the Mantua region, where he also associated with the military strategist and writer Macoio Lanteri. From 1550 he was a resident of Brescia, where he moved in the highest intellectual circles. Besides his two treatises on fortification and on surveying, Cataneo published *Avvertimenti et esamini intorno a quelle cose che si richiede a un bombardiero* (Brescia 1567, Venice 1582). Cataneo does not dedicate this book to anyone in particular but in his introduction mentions that it had been praised by Vespesiano (*sic*) Gonzaga and Giovan Battista Martinengo, both eminent military figures employed by the Spanish government in Milan.

The 1571 *Dell'arte militare libri tre* and the 1584 *Dell'arte militare libri cinque* include the two texts published in 1564, *Opera nuova di fortificare, offendere et difendere* and an appended text on firearms, and are illustrated largely with the 1564 blocks, although the large plate labelled "Alloggiamento campale" in the first edition at Harvard is not part of this copy. This illustration is significant because it is the only one that shows interest in the interior of the fortified enclosure. Otherwise Cataneo does not deal at all with the internal layout of the fortress or the fortified city.

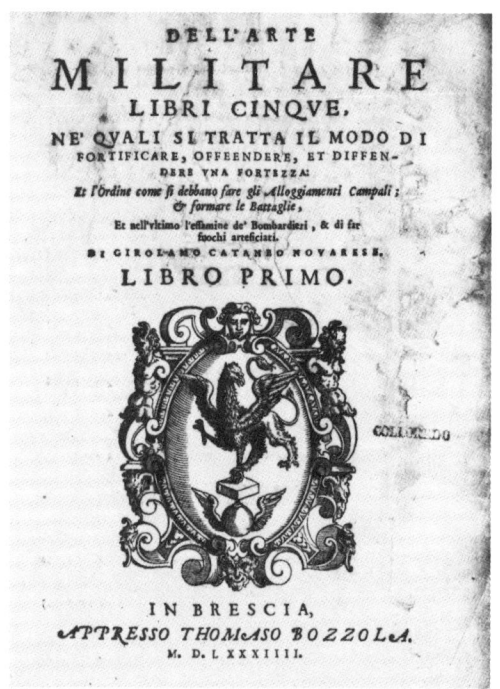

LIBRO PRIMO. CAP. QVINTO.
DECIMASESTA FIGVRA.

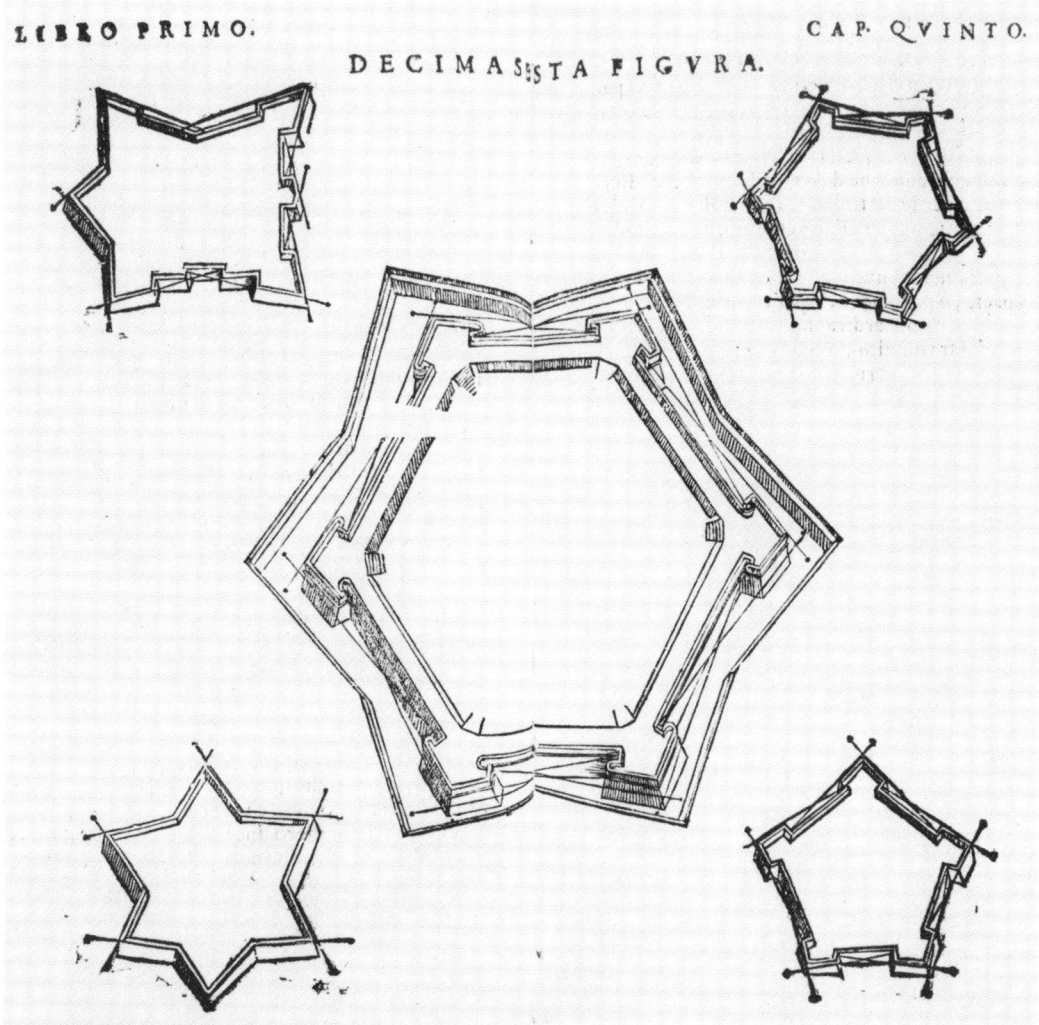

8 **Cataneo (or Cattaneo), Pietro,** 1510–1569.
I quattro primi libri di architettura.
Venice, sons of Aldo [Manuzio], 1554.

23.2 × 32.3 cm. Not bound.
ii + 54 leaves + ii. 43 woodcut plans of ideal fortified towns and plans and elevations of churches and palaces of which seven are full-page, 56 × 49 to 230 × 322 mm.
First Edition. Dedication to Enea Piccolomini; ten-year papal license and 15-year Venetian license to publish.
REF: HCL, Cockle, Wiebenson, Tafuri, Fortifications
COPIES: MH, BM, Bodleian (1554, 1567), BRT, NcD, CU, DAIA, MiU, NjP, DLC, MoU, WaU, CaBVaU, OrU, IU, OrCS, OKentC, GAT.

CATANEO WAS AN ARTIST, ARCHITECT and engineer from Siena. He was trained in the circle of Baldassare Peruzzi, with whom he worked closely, possibly superintending his construction projects. After his master's death in 1536 Cataneo was employed as fortifications architect and visited many outlying towns of the Sienese republic. His earliest publication is the *Pratiche delle due prime matematiche* (Venice, 1546).

This treatise was probably produced during a lull in the Sienese wars in 1553. As his illustrations of churches show, Cataneo was deeply indebted to the theoretical work of Francesco di Giorgio Martini (as seen in his drawings preserved at the Gabinetto delle Stampe e Disegni at the Uffizi in Florence) and to the published treatises of Sebastiano Serlio, both from Siena. The woodblocks from the 1554 publication were reused for the 1567 edition of *L'architettura.*

The four original parts of the *Quattro primi libri* deal with fortification, building materials, churches, palaces and private houses. The most original of the chapters is the one on military architecture. While illustrating his ideal city, with orthogonal street lay-out, regular fortification and overall star-shape, Cataneo seems to have drawn upon his experience in actual fortifications for the Sienese wars. Cataneo was the last of the Renaissance architectural treatise writers to consider both military and civil architecture together. Thus his work marks an important moment before the specialization of architects deepened.

Among the points he makes in the book on cities is that the capital of a kingdom should be at the center of its territory, foundations of towns and buildings should contain memorials to their builders, there should be a *pomoerium* between the city and its walls, the ambassadors of enemy or suspect nations should be lodged in buildings outside the city's walls, northern towns should have wide streets so that the sun can reach the buildings while the converse should hold for southern towns which should, however, have at least one wide and straight street so as to impress strangers, imitating the Roman Campo Marzio.

The illustrations in this work, especially the full-page ones, are dramatic and confident; many are presented in an "oblique" perspective unusual at the time but which was to become common.

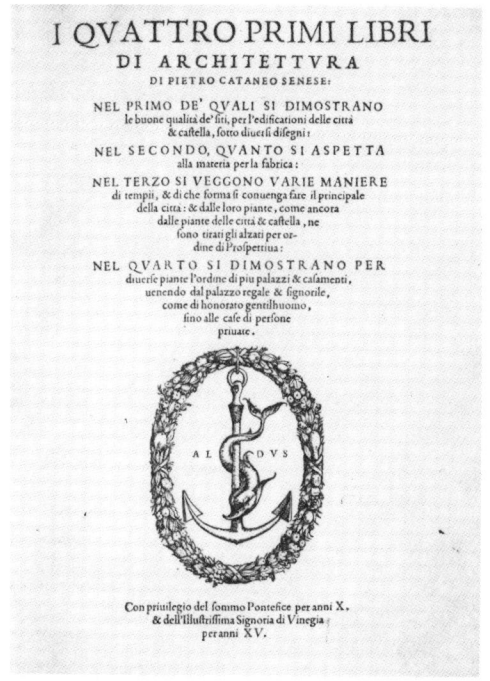 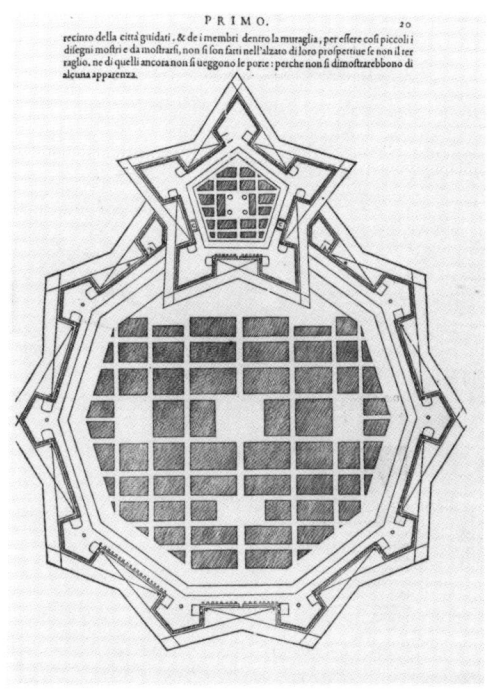

9 **Cellarius, Andreas.**
Architectura militaris.
Amsterdam, heirs of Jodocus Jansson, 1656.

22.5 × 34 cm. Contemporary vellum, blindstamped at center of covers, four ties, one removed, exlibris Theodosii Caroli Block 29 October 1799 inscribed inside upper end paper.
vi + 364 pages + xlv. Engraved title-page by I. v. Meurs, concordance of German, Flemish, French and Latin fortification terms, extensive calculations in text, numerous figures on 89 double-page engravings, 4 folded sheets, 3 identified plans of fortifications, 208 × 320 to 608 × 340 mm, of fortification studies and fortress plans, 13 double-page tables of dimensions. The tables and illustration sheets are bound as folded leaves in this copy.

Second Edition. The dedication to Christina, Queen of Sweden, and a separate one to five of her Swedish generals, are not dated, but first edition was published in Amsterdam in 1645.

REF: Michaud
COPIES: BN, HAB (1645), MiU-C (1645), DLC, NjP (1645).

A GEOGRAPHER, COSMOGRAPHER AND mathematician, Cellarius was the rector of the College at Hoorn, Holland. His *Description de Pologne et de Lithuanie* (Amsterdam, 1652, 1659) was translated into Dutch in 1660, while the *Harmonia macrocosmica, seu Atlas universalis et novus totius universi creati* (Amsterdam, 1661) was joined to Blaeu's atlas in a publication of 1708.

Divided into four books, this weighty treatise deals with regular fortification, outworks, irregular fortification, and the attack and defense of a fortified city or fortress. His signature—Palatinus exul—hints that he was a native of the Palatinate.

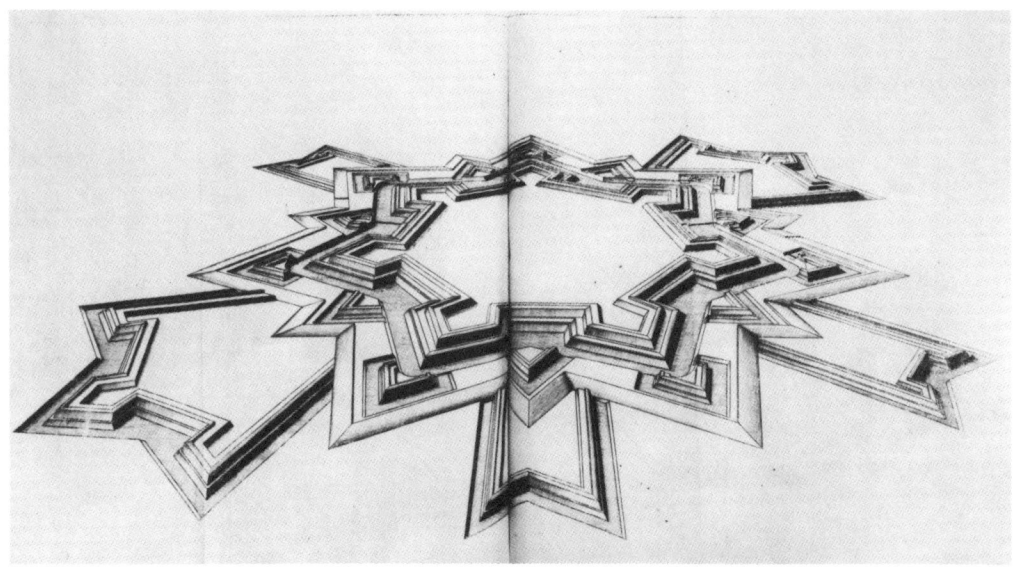

10 **Coehoorn, Minno** (or Menno), Baron van, 1641–1704.
The New Method of Fortification.
London, Daniel Midwinter, 1725.

19.5 × 31.9 cm. Brown calf, blindstamped, rebacked, repaired lower cover, ex-libris Charles E. Dashwood, Esq., Wherstead Park.
xx + 181 pages + i. 3 full-page engravings of fortification sections, 13 folded engravings of fortress plans, 190 × 319 to 450 × 514 mm.
Second English Edition. Dedicated to the Prince [of Wales] by Thomas Savery, translator.
REF: Architekt und Ingenieur, Michaud, Duffy
COPIES: MH (French version, Wesel 1706), BN (Leeuwarden 1685, French, The Hague 1706, 1711, 1741, 1743; Wesel 1706; Moscau 1706), HAB (Dutch, Leeuwarden 1685; French, Wesel 1706, The Hague 1706; German, Wesel 1708), T (1705), DNW (French, The Hague 1711), MiU (French, Wesel 1706), DN (French, Wesel 1706), Vi (London 1705), MnU (London 1705; French, The Hague 1741), N (London 1705), DFo (London 1705), WU (Leeuwarden 1702), PPAmP (Leeuwarden 1685), PPULC (Leeuwarden 1685), ViU (French, The Hague 1741), NWM (French, The Hague 1741).

A MILITARY ARCHITECT AND MATHEmatician, Coehoorn was born into a distinguished family; his officer father inspired him with interest in military science from his childhood. He entered military service at 16 after receiving a thorough education in mathematics. Coehoorn, the "Dutch Vauban," became General of Artillery (1697) and Lieutenant General of the Infantry of Holland (1695), and Governor of Flanders and of the fortresses on the Scheldt. His masterpiece is the fortification of Bergen-op-Zoom in preparation for the Franco-Dutch Wars where he opposed Vauban, especially at Namur, which the French and the Dutch lost and regained in turn. Coevorden is the best example of his fortification theory put into practice.

This is a splendidly illustrated posthumous edition of Coehoorn's treatise on fortification, first published in 1685 in Dutch as *Nieuwe Vestingbouw*. Divided into nine chapters, it concentrates on the "method of strengthening the interior space of the French Royal Hexagon," with separate chapters dedicated to the heptagon and the octagon. The French method was chosen because "we have found no better among the many designs made by the lovers of this Art." Further Coehoorn says that "the plan of it was given me some years ago in the name of a great engineer" suggesting that Vauban's improvements were known to him. While recognizing the French manner of fortification as the best and the most popular throughout Europe, Coehoorn contends that the invention of this method should be credited to Specklin, who published it in 1589. Echoing Pagan, he marvels at the slow evolution of fortification science ("Is it not then a wonderful thing, that in a whole Age there should be so small Improvements made in the Art of Fortification?") and proceeds to criticize the false improvements made by Marolois and Freitag which had held sway most of the seventeenth century. Savery (1650–1717), the translator of this edition, was employed as a military engineer at Hampton Court. He is better known as the inventor of the steam engine.

Fig: O.

A Scale of 120 Rynland Rod

11 Croce, Flaminio.
Theatro militare.
Antwerp, Henry Aerts, 1617.

16.4 × 24.2 cm. Brown calf boards, armorial of Maria-Augustus von Sultzbach on upper cover, gold-stamped spine, title on red morocco panel, ex-libris Gustave Charles Antoine Marie van Havre, and of J. W. Six inside upper cover, marbled edges.
xxiv + 343 pages. Introduction, table of contents, 15 engraved illustrations, 158 × 150 to 148 × 230 mm.
Second Edition, enlarged. Dedication to Johann Jakob, Count of Bronckhorst, Secret Councillor of Leopold Archduke of Austria, is dated 14 January 1617. License to publish given by the Archbishop of Antwerp 26 September 1616; archducal license to publish dated 10 October 1616 from Brussels.

REF: Cockle
COPIES: BM (1613), NN (1613), MiU-C (1613), MnU, NcD.

Croce was an aristocrat from Milan who had served many years in the Spanish army in Flanders. His *Essercitio della cavalleria et d'altre materie* was published in Antwerp in 1625, and reprinted in 1628 and 1629.

This work of military art is divided into four *Discorsi*. Questions of fortification are touched upon in the first part where the role of the governor, the provisioning of the fortress and its physical defenses are discussed in terms of overall strategy.

THEATRO MILITARE
DEL
CAPITANO FLAMINIO
DELLA CROCE,
Gentil'huomo Milanese.

La Seconda volta dato all' Impressione con l'aggiunta di molte figure, molti Capitoli nuoui, & gli altri tutti ampliati.

DEDICATO
All'Ill.mo Sig. Gio: Iacomo, Conte di Bronchorst, Barone di Bateborch, & Anholt, &c.

IN ANVERSA,
Appresso HENRICO AERTSSIO.
M. DC. XVII.
Con Priuilegio.

12 **Deville, Antoine,** 1596–1657.

Les fortifications ou l'ingenieur parfait.
Amsterdam, 1672 (on added title-page A. Wolfgang, H. and T. Boom, 1675).

11.2 × 18.5 cm. Contemporary brown calf, goldstamped spine with title on beige panel.
xvi + 381 + xi. Engraved title page [Amsterdam, Abraham Wolfgang, Henry and Theodore Boom, 1675], introduction, table of contents, 13 illustrations of fortifications engraved by the author folded-in, 298 × 184 to 427 × 184 mm, 40 illustrations in text engraved by the author, 73 × 60 to 74 × 136 mm, subject index.

Sixth Edition.

REF: Cockle, Manno, DBF
COPIES: BM (1640, 1672), BN (1628, 1636), Bibliothèque du Génie-Paris (1628), MiU-C (1628, 1666) MH (1666), MnU.

ANTOINE DEVILLE [OR DE VILLE] WAS AN engineer and military architect. Member of a distinguished French military family (his brother served the prince of Carignano and the duke of Savoy), Deville was awarded the order of S. Maurizio and S. Lazaro by Carlo Emanuele I, duke of Savoy. He authored a number of publications which document his extensive experience in siege warfare. These include *Les Fortifications* (Lyon 1629, Paris 1636, Amsterdam 1672), *Obsidio Corbeiensis* (Paris 1637) and *Le Siège de Hesdin* (Lyon 1639) all three with illustrations. A contemporary and rival of Blaise Pagan, he had studied military architecture using the treatise by Jean Errard; as an expert with mines he participated in many sieges. While in the service of the Republic of Venice from 1630 to 1635 he fortified Pola. Upon his return to France he entered the service of Cardinal Richelieu. In 1636 he was in Flanders with the Count of Soissons, in 1637 at the sieges of Corbie and Landrecy, and in 1639 at the siege of Hesdin. His numerous military activities and travels were documented in his abundant publications. These include *Pictomachia veneta seu de pugna venetorum in Ponte* (Venice 1633), *Descriptio portus et urbis Polae antiquitatem* (Venice 1633), and *Le Siège de Landrecy* (1637).

His treatise is divided into three parts: the design, the attack and the defense of fortresses. He insists that the true talent of the engineer will be best challenged by irregular fortifications since ready-made principles will not be applicable. In his approach, attack and defense are only aspects of the same complex fortress. Among his contemporaries he was considered as the best representative of the French method of fortification and juxtaposed to Marolois' Dutch defense. His claim that his designs are all based on his own experience in war ("Je t'asseure pourtant, amy lecteur, que je n'ay rien escrit, que mon frère [Serjeant major du Régiment de Monsieur le Prince Thomas] ou moy, n'ayons veu, ou pratiqué: l'experience de ceux qui s'en serviront fera connoistre la verité") is presented in the preface to the treatise.

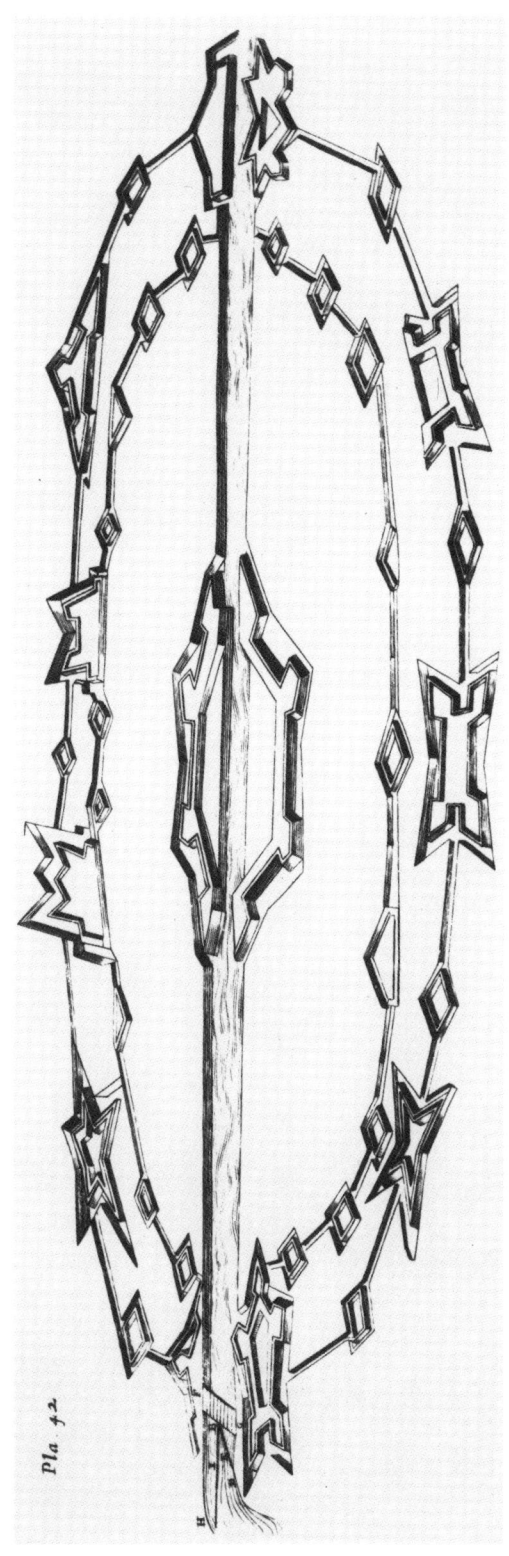

Pla. 42

13   **Deville, Antoine,** 1596–1657.
De la Charge des gouverneurs des places.
Paris, Matthieu Guillemot, 1639.

19.6 × 29.4 cm. Modern cloth-binding on boards, ex-libris "Nemo me impune lacessit" Through. xii + 292 pages + viii. Engraved title-page by H. David, anagram of the author's name and epigram by Billon and dedication by Des Maretz ("Toute fois quel progrès? puisque tu fais entendre / Si l'on suit tes leçons / Et qu'on peut prendre tout, et qu'on ne peut rien prendre"), preface, 7 full-page engravings, 168 × 358 mm, subject index.

First Edition. Royal license to publish and sell for seven years from publication date, 25 May 1639. Dedicated to Cardinal Duke of Richelieu.

REF: Cockle, DBF
COPIES: BM, BN, Bibliothèque du Génie-Paris, DCL, NjP (Paris and Leiden 1640), PU (Paris and Leiden 1640), WU (Paris 1666).

The Newberry owns also a copy of the *De la charge* Paris 1640 edition (brown worn calf) and a copy of the Rouen 1666 edition (new cloth binding, ex-libris Robert Keith of Craig, Esq. on verso of title-page and Bibliotheca Lindesiana inside upper cover).

14 Dilich, Wilhelm, 1571/72–1655.
Peribologia oder Bericht von Vestungs gebauen.
Frankfurt am Main, Anthonio Hummenn, 1640.

18.6 × 28.8 cm. Brown half calf, goldstamped decorations and title on red morocco panel, marbled paper boards.
163 pages + xiii. Backed titlepage drawn by Dilich and engraved by S. Furck, eight engraved part-title pages, two subject indices, 410 Roman-numbered illustrations, 140 × 86 to 364 × 288 mm, of these 73 are double-page illustrations folded-in, representing fortress plans, elevations of gates, sections through fortification trace and camp layouts, engraved plan of Strasbourg, 644 × 470 mm.

First Edition. Dedication to Johann Georg Duke of Saxony is dated 1 February 1640.

REF: Cockle, Marini, Architekt und Ingenieur
COPIES: BM (1640, Frankfurt 1689), BN (Latin version, Frankfurt 1641), HAB, NNC, MnU, NjP.

THE BEST-KNOWN GERMAN WRITER ON fortification of his time, Dilich had studied at Cassel and at the University of Marburg. He spent long periods of time in Holland and then became the historian, geographer and architect of Maurice, Elector of Saxony (joint dedicatee of the Latin edition of 1641); poems praising his treatise address him also as a mathematician.

This treatise is distinguished by the ferocity of the illustrations which attempt to create an aesthetic of fear, elaborated in the external form of the fortification, especially the casemates, the gun embrasures of the parapet and the gates. His outlines for fortified city plans are somewhat influenced by Francesco de' Marchi's, but have even more extensive outworks; Dilich clearly supported a fortification system based on the "tenaille," a predilection partially announced by his title (Peribologia = study of the circumference [of a sanctuary]). Nonetheless, many of his fortified city plans contain elaborate proposals for urban compositions of streets, squares and building lots.

CI.

CCLXXVI.
Jo: Wilh: Dilich.
Ingen:

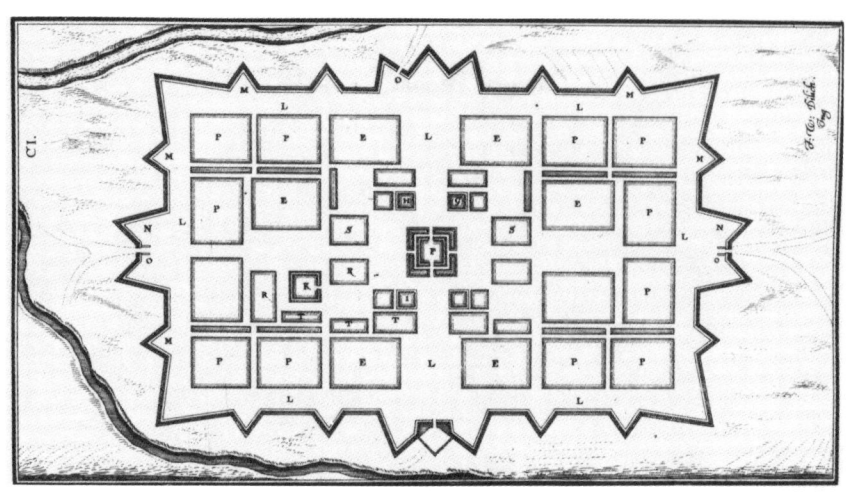

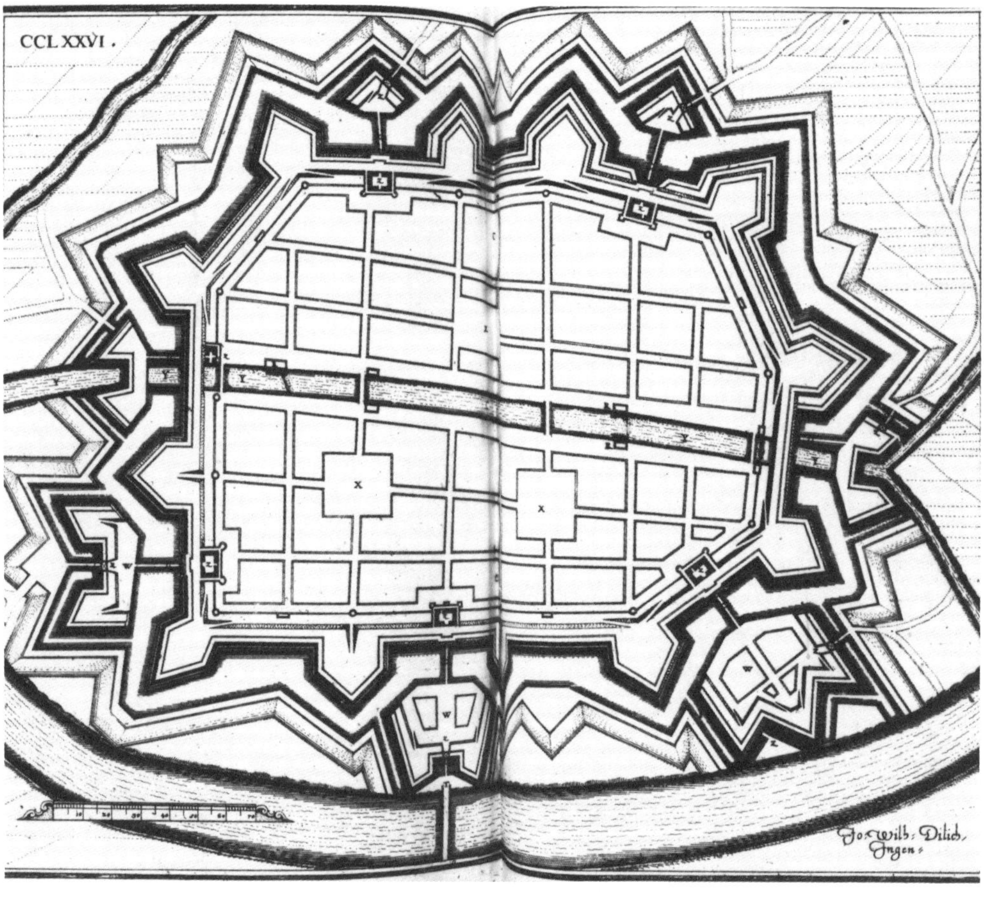

15 **Dögen, Matthias,** 1605–1672.
Architectura militaris moderna.
Amsterdam, Ludovic Elzevir, 1647.

20.2 × 31.2 cm. Contemporary torn, stained limp vellum, ex-libris John Cope, Bart. inside upper cover.
viii + 504 pages + xxii. Engraved title-page by T. Matham, instruction to binder for location of illustrations, 29 engraved figures labelled in alphabetical order at bottom right, 41 engraved plans of identified cities, 368 × 312 mm, each is a double-page, subject index.

First Edition. Dedication to Frederick Wilhelm, Duke of Brandenburg.

REF: Architekt und Ingenieur
COPIES: HAB (French and German 1648; Latin 1647), BN (French 1648), MCM (French 1648), MiU (1648), DLC, OCL, PU, MdBP, McD, NNC, IC, InNd (French 1648), NjP (French 1648), MH (French 1648), PU (French 1648), NcGW (French 1648).

DÖGEN STUDIED IN LEIDEN AND HELD AN office in the Admiralty Headquarters in Amsterdam, but although he was named member of the Council of Brandenburg and its representative in Holland, nothing is known of his activity as a fortification architect.

This treatise was published almost simultaneously in three languages, Latin (1647), French and German (1648), by the same publisher in Amsterdam, Elzevir. The French edition was dedicated to William Prince of Orange and it is also an elegy for Maurice of Orange, who is called "the father of fortification" since he had turned the United Provinces into a single fortress; the German version does not carry a dedication. The French (viii + 547 pages + i, 19.4 × 29.1 cm) and German (viii + 475 pages + i, 20. × 29.8 cm, Bibliothek des Heeresmuseum title-leaf verso) versions—also in the Newberry Wing collection—have the same number of illustrations; in the French version the double-page is folded in rather than bound. The German version is bound in contemporary worn limp vellum, while the French example is bound in brown calf (gold-stamped decoration and title on the spine, ex-libris Henry Probasco, 1 December 1890). In the German version the text is preceded by a valuable con-

28

cordance in German, Latin and French of fortification terminology.

According to Dögen military architecture is divided into two parts: *Hercotechnique,* that is, the method of fortifying places and *Areotechnique,* that is, the methods by which one takes or defends a place. He discusses the terminology of the elements of fortification through three modes of representation: ichnography, orthography and scenography, that is, plan, section and view. He adopts the same approach—Vitruvian in its inspiration—when he writes about the economy of fortification, and when he provides rules and maxims for the relationship between the parts of fortification. Among ancient authors he quotes are Vegetius, Livy, Salust and Pliny; he seems equally familiar with contemporary military architects and their major works, offering praise when appropriate (to Paciotto, for instance, and his design of the citadel in Antwerp, p. 296).

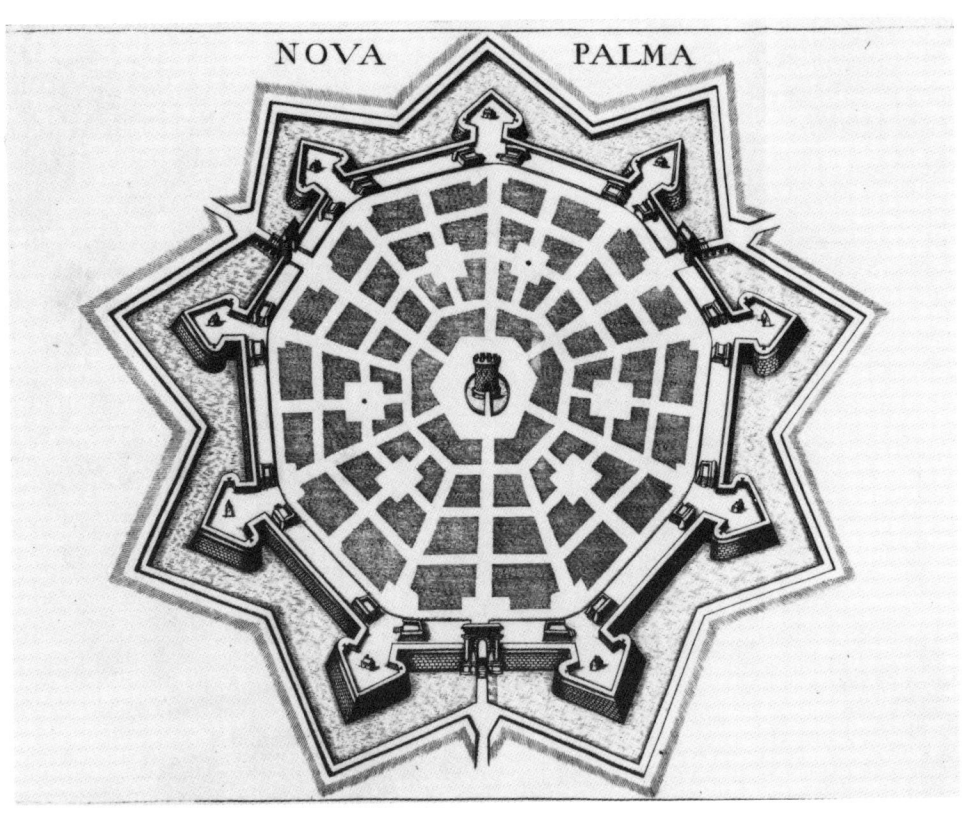

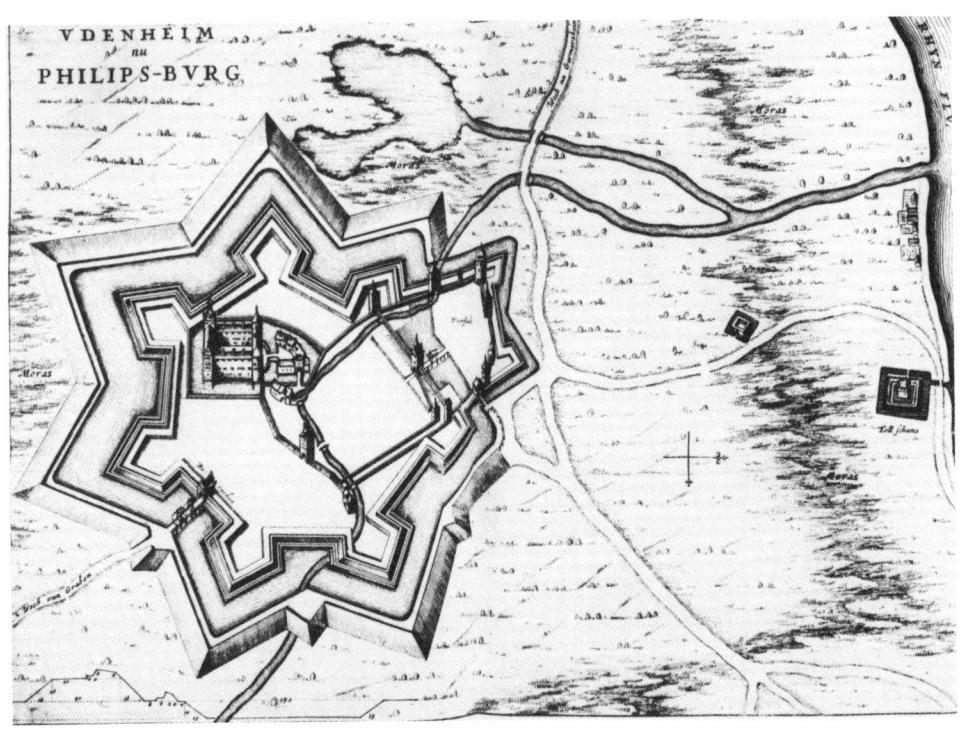

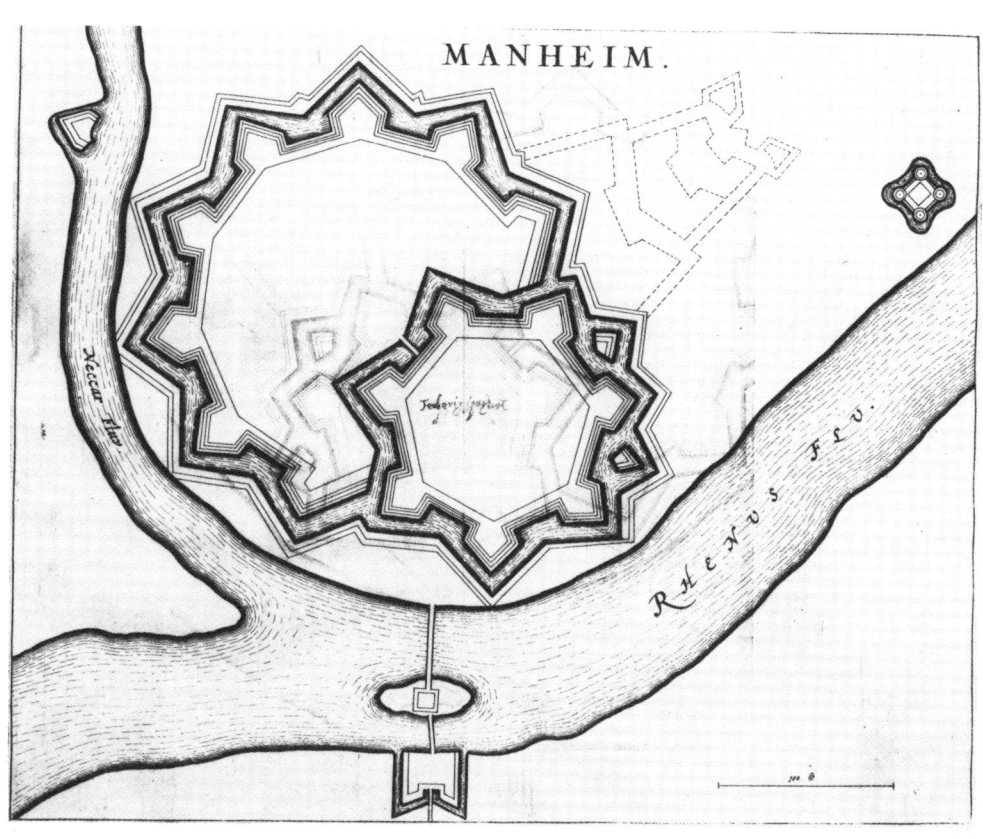

16 Dubreuil, Jean, Sieur de Bitainvieu, 1602–1670.
L'Art universel des fortifications françoises, hollandoises, espagnoles, italiennes, et composées.
Paris, Jacques Dubreuil and François Eschart, 1674.

18.4 × 25 cm. Contemporary brown calf, goldstamped decorations and title on spine, marginalia in German, H. Zöllner inside upper cover.
xxxii + 361 pages + iii. Half-title, engraved frontispiece and part title-pages for parts I, III and IV by Le Pautre, 122 full-page illustrations paginated with the text, 122 × 176 mm.
Second Edition, corrected and enlarged. Published with royal license dated 27 October 1664, first edition printed 28 February 1665.

REF: DBF, Wiebenson
COPIES: BN (1665), NN, NjP, ViW, MnU (1665, 1674), MiU, NNH, DNW (1665).

SON OF A BOOKSELLER, DUBREUIL JOINED the Jesuits in 1624; he then studied and practiced architecture while living for several years in Rome. Among his publications are the famed *La perspective pratique, nécessaire à tous peintres, graveurs, sculpteurs, architectes* (Paris 1642–48, 3 vols.) which became known as the "Jesuit perspective," augmented with a *Traité de la perspective militaire ou méthode pour élever sur des plans géométraux,* and *Diverses méthodes universelles et nouvelles, en tout ou en partie, pour faire des perspectives* (1642).

Divided into six parts, the book's major topics are the terminology of fortification, fortification of a regular polygon, fortification of irregular polygons, the design of fortification on paper, the art of attack and defense of a fortress, and a comparison between the "national" fortification methods, French, Italian, Spanish and Dutch. The work is addressed to young noblemen studying in military academies; the education of the aristocracy in the art of war is illustrated in one of Le Pautre's plates, which shows Pallas lecturing to a group of armored and bewigged young men.

Ieunes Heros que Mars instruit a la Victoire, Afin de meriter le haut point de la Gloire;
Il faut sçauoir, et Vaincre, et dresser vn Rempart, Et Pallas vous en monstre elle même icy l'ART.

17 **Dürer, Albrecht,** 1471–1528.
Etliche Underricht zu befestigung der Stett, Schloss, und Flecken.
Nuremberg, [Hieronymus Andreae], 1527.

20.8 × 28.8 cm. Beige half calf, speckled tan paper boards, goldstamped spine with Habsburg coat-of-arms. This copy is bound with two other works by Dürer: *Unterweysung der messung* (Nuremberg, 1525) and [*Hierinn sind begriffen*] *vier Bücher von menschlicher Proportion* (Nuremberg, 1528). 26 leaves. Title-page with arms of Ferdinand I, King of Hungary and Bohemia, 22 woodcut figures, 168 × 37 to 410 × 280 mm.

First Edition. Dedication to Ferdinand I, King of Hungary and Bohemia.

REF: Cockle, Architekt und Ingenieur, Waetzoldt
COPIES: BM, HAB (two copies), WU, DLC, NN, CtY, KU, MWiW-C, NIC, OC, NjP.

KNOWN AS THE FOREMOST GERMAN artist—painter, engraver, and draftsman—of the sixteenth century, Dürer also made significant theoretical contributions to such contemporary concerns as perspective, proportions, calligraphy and fortifications. Writing in the vernacular, Dürer, like Luther, had to create a German language of his own. His construction of the pentagon stimulated the imaginations of Cardano, Tartaglia, Galileo and Kepler, and Pietro Antonio Cataldi dedicated a monograph to the "Modo di formare un pentagono . . . descritto da Alberto Durero" (Bologna 1570). Dürer's ideas about city-planning are drawn from German medieval tradition and the principles of the Italian Renaissance. They reveal his familiarity with such modern theoreticians as Leon Battista Alberti and Francesco de Giorgio Martini, and the inspiration he found in the proposals of Filarete, in the circular strongholds that he saw on his first trip to Italy, and occasionally in Leonardo's studies. Dürer endeavored to raise "usage" to the level of art by formulating a scientific theory of the art of fortification based on the practice of military engineers in Germany and Italy; his legacy is the ennoblement of the artistic profession by giving it a share in the responsibility for cultural and political life.

Dürer's treatise covers four subjects: 3 different ways of building a bastion; a project for a blockhouse; the planning of an ideal city; suggestions for the strengthening of existing fortifications. In his design he mixes reality and fancy, as usual, and his gigantic bastions are proportioned without consideration for finances. He provided the first German example of the polygonal front in his fortified town and castle; his suggestions include a radius of one mile of open land around the fortification, double ramparts, and main and secondary moats.

But his scheme for the development of the ideal town is the most interesting part of his treatise. Although defense dominates the whole plan, with its regular layout derived from the Roman "castrum" and hygienic conceptions borrowed from Antiquity, he distributes the various functions carefully in certain quarters and streets thus mingling Gothic-German traditions and Italin neo-classical ideals. The communal center of the city—market place and townhall—is before the eastern gate, the eastern corner is occupied by the church and vicarage, the food industry is the north-east corner—butchers, bakers and brewers—while the south-west corner is given to the arsenal, granaries, timber warehouses and houses for artisans. Although his ideal city has straight, broad streets he admits that other ideas on town-planning have an equal claim to recognition, just as beside "ideal" men (Apollo) other human beings of different proportions have an equal right to existence. The only possible parallel to Dürer's ideal town is the celebrated Fuggerei (one of the earliest "slum-clearing" projects in history) in Augsburg, begun in 1519 and sponsored by the Fugger with whom Dürer had close connections. Dürer's dreams were realized by Speckle in 1589, but his study of the theory of military architecture is a presage of a new development which was to appear in the German Baroque, that is, military engineering as a basis of monumental architecture, with military engineers and artillery officers, such as Speckle, Dientzenhofer and Neumann as masters of sacred and civil architecture.

18 **Dürer, Albrecht,** 1471–1528.
De urbibus, arcibus, castellisque condendis, ac muniendis rationes aliquot, praesenti bellorum necessitati accommodatissime.
Paris, Christian Wechel, 1535.

20 × 30.6 cm. Brown calf on wood boards, blindstamped, metal corners, two metal clasps and hooks removed, manuscript vellum paste-down endpapers. This copy is bound with *Euclidis megarensis Mathematici clarissimi elementorum geometricorum* (Basel, Johann Hervag, 1537) and Dürer's *Geometricarum libris* (Paris, Christian Wechel, 1535).
38 leaves. 23 woodcut figures, 167 × 35 to 408 × 306 mm.

First Latin Edition.

REF: Cockle
COPIES: BM, Bodleian, HAB.

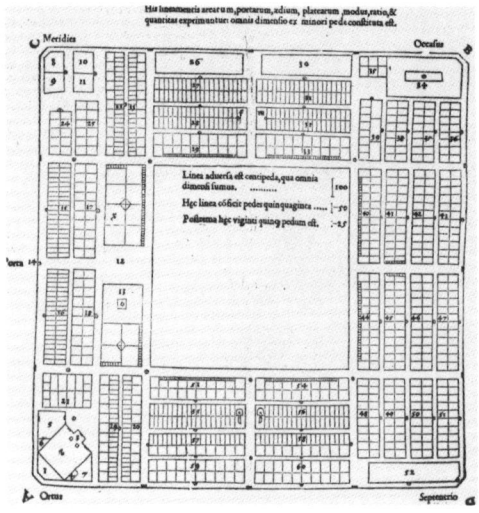

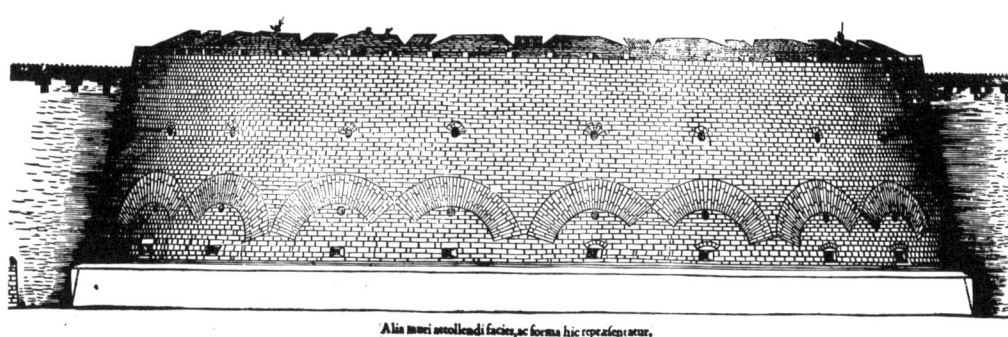

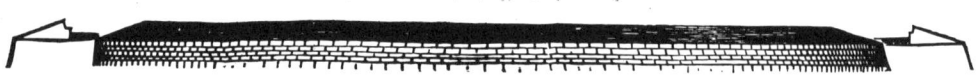

19 Errard, Jean, de Bar-le-Duc, 1554–1610.
La Fortification reduicte en art et demonstrée.
Frankfurt, the heirs of Theodore de Bry and Wolfgang Richter, 1604.

19.3 × 28.7 cm. Brown calf, worn at spine and corners, goldstamped decoration on boards and spine, title and date goldstamped on spine, Henry-Joseph Pierson inside marbled free endpaper.
viii + 77 pages. Engraved title-page, preface dedicated to French nobility, 38 engraved illustrations of fortifications numbered in arabic (except Fig. XXVI), each illustration is double-page, 226 × 80 to 228 × 232 mm, many line drawings in text.
Pirated edition. Dedication to Henry IV King of France is dated January 1600.

REF: Promis, Cockle, Architekt und Ingenieur, Manno, DBF, Lallemand
COPIES: HAB, BM (Paris 1604, 1620), BN (Paris 1600, 1620), MiU (Paris 1604, 1620), CtY (Frankfurt 1604; Paris 1620), MH (Frankfurt 1604; Paris 1620, 1622), NN (Frankfurt 1604; Paris 1620), DFo (Paris 1604), NNC (Paris 1620), NWM (Paris 1620), ViM (Paris 1620), MnW (Frankfurt 1617), NjP (Frankfurt 1617).

BORN INTO A FAMILY FROM LORRAINE raised to the nobility in 1470 by King René, Errard studied mathematics and fortification in Italy. Upon his return he entered the service of Charles III, Duke of Lorraine, to whom he dedicated his *Premier livre des instruments mathématiques et mécaniques* (Nancy 1584). In royal service he obtained in 1591 the license to mint coin. In 1594 he published *La géométrie et pratique d'icelle* and then spent two decades in Henry IV's wars. In 1595 he was in Sedan, in 1596 at Calais, in 1597 at the siege of Amiens and in Guise, in 1598 he built the new citadel of Amiens; in 1603 he was in Burgundy, in 1606 in Reims and 1609 in Metz serving the king on military missions. He was personally ennobled in 1599 and became the "ingénieur ordinaire" of the King in Picardy and Ile-de-France.

This French version printed in Germany follows two known versions printed in Paris (1600 and 1604), which may in turn have been preceded by an edition of 1594, considered the first edition by Promis. There are two indications that the Frankfurt edition was pirated: the note to the reader ("quelques gens de bien considerans que les exemplaires de ce livre imprimé a Paris, estoient rare dans ces quartiers, et de tres cher pris, nous ont induicts a le faire

veoire icy non seulement en François, mais aussi en Allemand, et cé en forme plus petite" [sic]), and the subtitle of the posthumous edition of 1620 ("revue corigée et augmentée par A. Errard son nepveu aussi ingenieur ordinaire du Roy suivant les memoires de l'auteur contre les grandes erreurs de l'impression contrefaicte en Allemaigne"). The title page of the Paris 1600 edition is the same design, but the cannon-order columns are both labelled M. de Béthune. Errard insists on the importance of theory—"practice is blind without theory and theory is awkward (*manchotte*) without practice"—in his preface to the French nobility, and claims that his book is the first that goes beyond mere mechanical discourse.

This work—which aims to clarify and render honorable a difficult "human science" (p. 77)—is divided into four books. Assuming that the reader is familiar with the nomenclature of fortification, Errard begins with a series of axioms, followed by a discussion of the army, the defense and the assault of fortresses and the qualities required of an engineer. Book two, dedicated to Maximilien de Béthune, superintendant of French fortifications and artillery, is about the outline of the fortress ranging from a hexagon to a 24-sided polygon. Books three and four deal with irregular and dominated fortresses respectively.

Jean Errard was one of the most important members of the Corps du Génie, founded during Henry IV's reign by his Prime Minister Sully. Errard's most formidable competitor was Claude Chastillon, a prolific engineer and cartographer. Errard's particular interest was in the ideal-city plan, which eventually he designed in versions ranging from a hexagon to a 24-sided regular polygon. He showed how to mask the bastions through the disposition of the curtain walls and invented the "cavalier," a platform raised above the bastion or the curtain wall from which the battery could sweep the countryside. He seems to have been responsible for many urban interventions made during Henri's brief reign. His is the first systematic treatise on fortification in French.

20 **Errard, Jean, de Bar-Le-Duc,** 1554–1610.

Fortificatio, Das ist: Künstliche und wolgegründte Demonstration und Erweisung wie und welcher Gestalt gute Festungen anzuordnen.

Frankfurt am Main, the heirs of Dietrich de Bry and Wolfgang Richter 1604.

19.9 × 29.8 cm. Contemporary vellum, worn at corners, stained on lower board, blindstamped, exlibris Wedel, Baron Jarlsberg.

viii + 71 pages + 15 (11–14 missing). Engraved title-page, 37 engraved illustrations of fortification plans and views, numbered in arabic (except Fig. XXVI), 221 × 59 to 226 × 232 mm, the illustrations are folded in.

First pirated German Edition. Dedication to Henry Julius, Bishop of Halberstadt and Prince of Braunschweig, is dated 1 March 1604.

REF: Architekt und Ingenieur, Cockle, Manno
COPIES: HAB.

The German edition in the Newberry collection is bound with Johann Jacob von Wallhausen, *Camera militaris oder Kriegskunst Schatzkammer* (Frankfurt 1621, xxviii + 244 pages + xxxviii), dedicated to Maximilian, Duke of Bavaria.

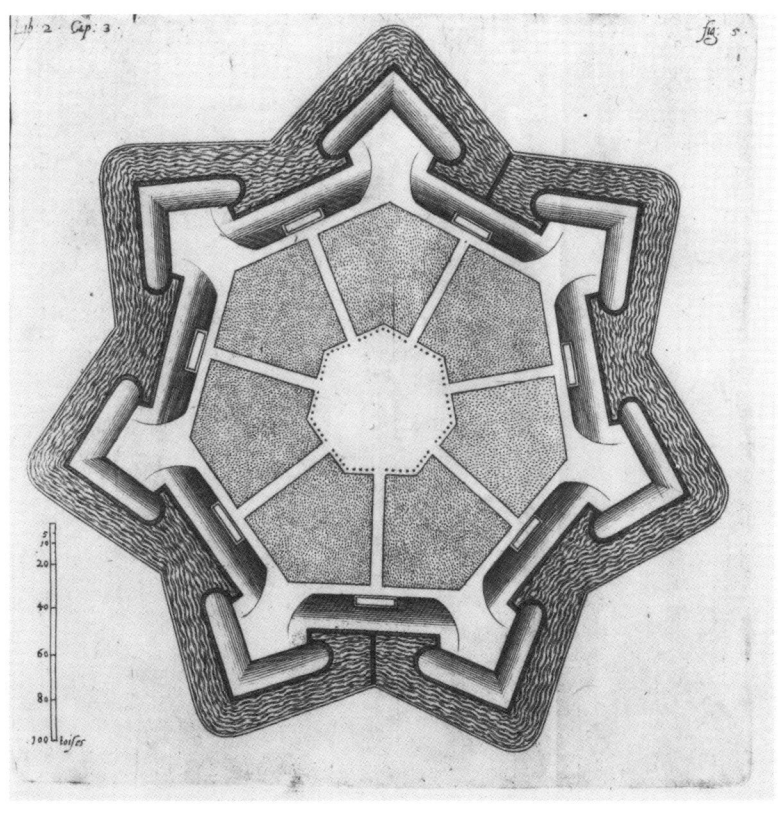

39

21 **Fer, Nicolas de,** 1646–1720.
Les forces de l'Europe, ou description des principales villes avec leurs fortifications.
Paris, Nicolas de Fer, 1695–96.

37.5 × 28.1 cm. Contemporary brown calf, blindstamped decorations, goldstamped title on brown morocco panel on spine.
Unpaginated. Two part title-pages, eight part tables of contents, 184 engraved figures, views of cities, siege and city plans, 183 × 122 to 520 × 358 mm, first 8 sheets of figures accompanied by separate explanatory legends.
First Edition. Dedication to Duke of Burgundy, royal license to publish dated 1693.

REF: Pastoureau
COPIES: ICN (1705; 1723 [published by J. F. Benard, de Fer's son-in-law]), BN (1723), DLC (1696–97), NN (1695), OU (1690–95), MN (1690–95), NNC (1693–95), NNC-A (1693–96), MiU (1690–95), MH (1695–96), NjP (1695–96).

Nicolas de Fer held the title of geographer at Louis XIV's court. He was one of the most prolific, though not innovative, publishers of plans of European cities in the late seventeenth century. This work is only marginally a treatise on military architecture or fortification strategy. Rather than attempting to teach the design and construction of fortification de Fer hopes to instruct by providing an ever-growing and exhaustive series of examples among realized fortifications. The illustrations often include a verbal description and a legend with the principal buildings, as well as labels directly on the plan; each city plan includes both fortifications and street layout. The two signed engravers are van Loon and Schoonebeek, while the title-page proclaims that the plans were surveyed and drawn by the best engineers, especially French ones under the command of Vauban. Despite the title, the emphasis is upon French towns; the 1705 and 1723 editions begin with a table of contents arranged by country with part title-pages removed.

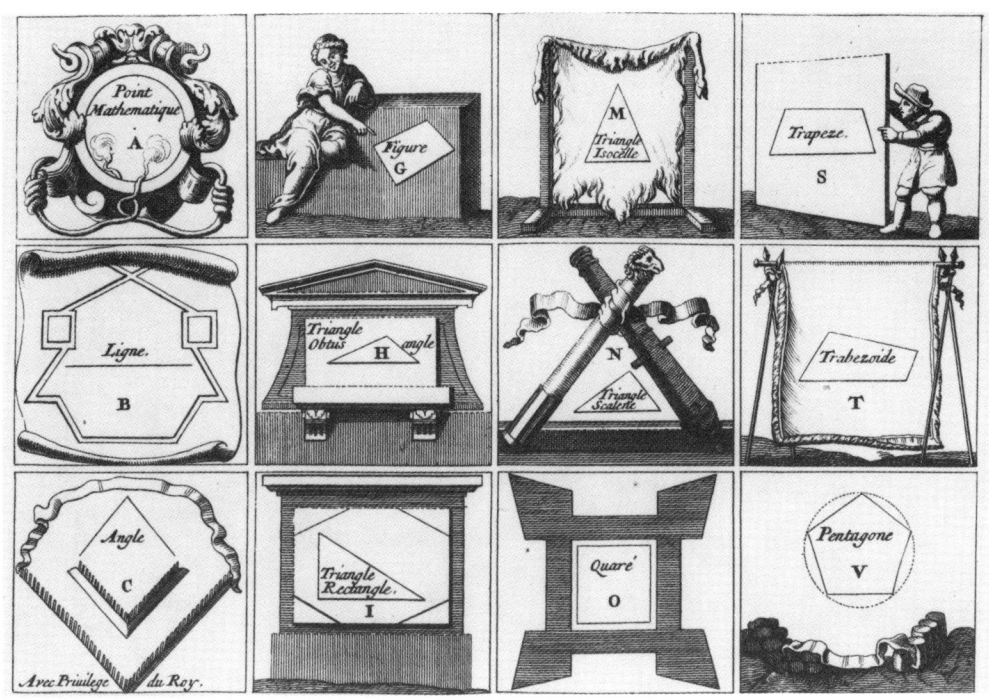

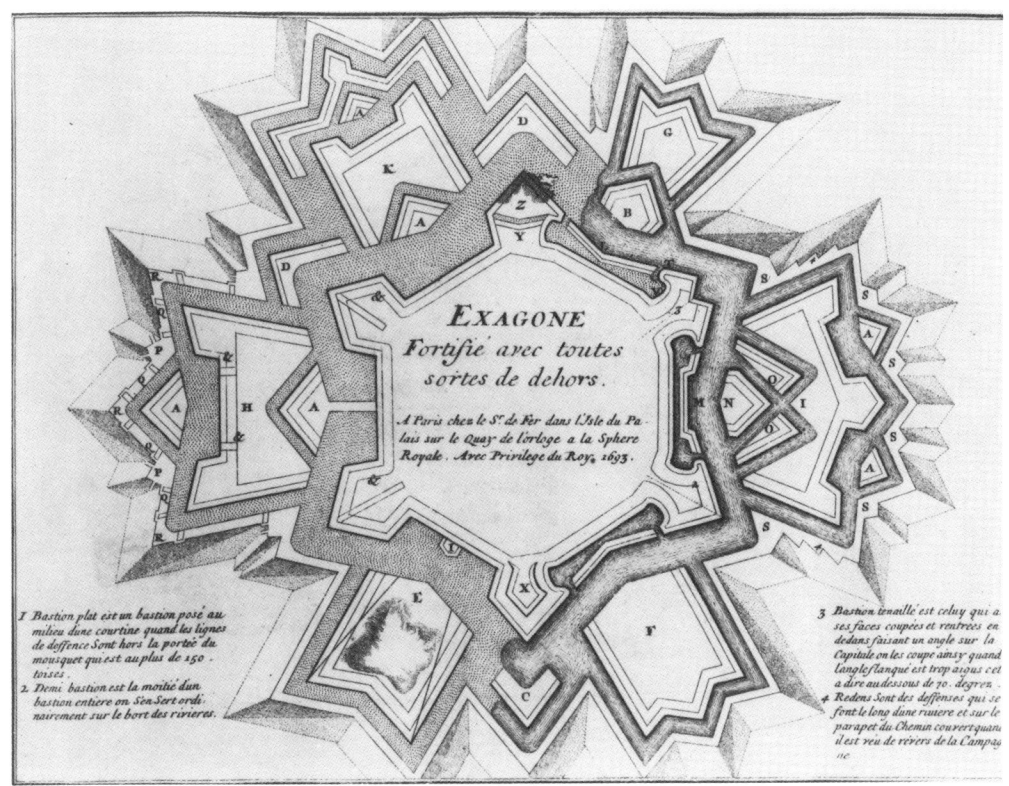

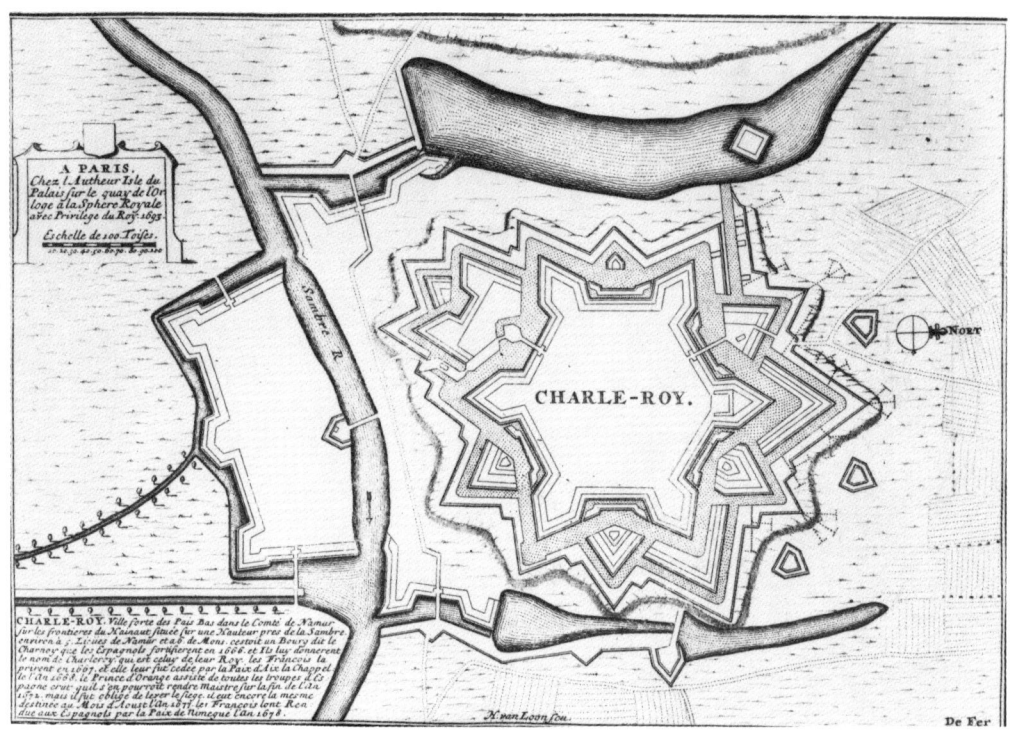

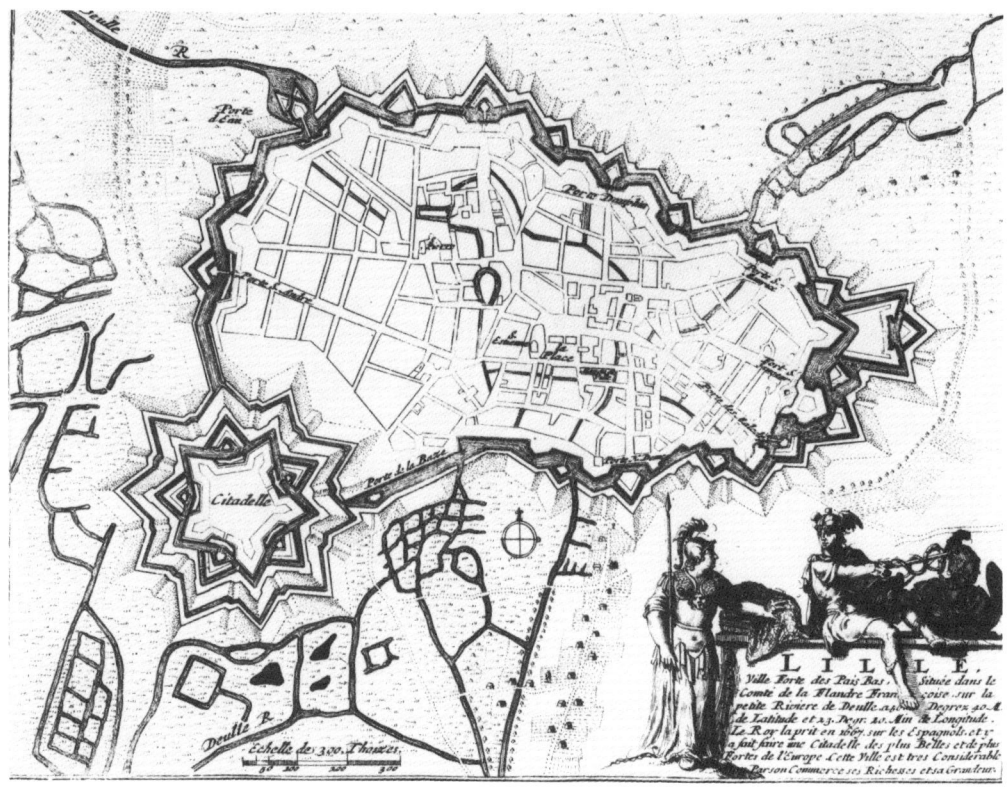

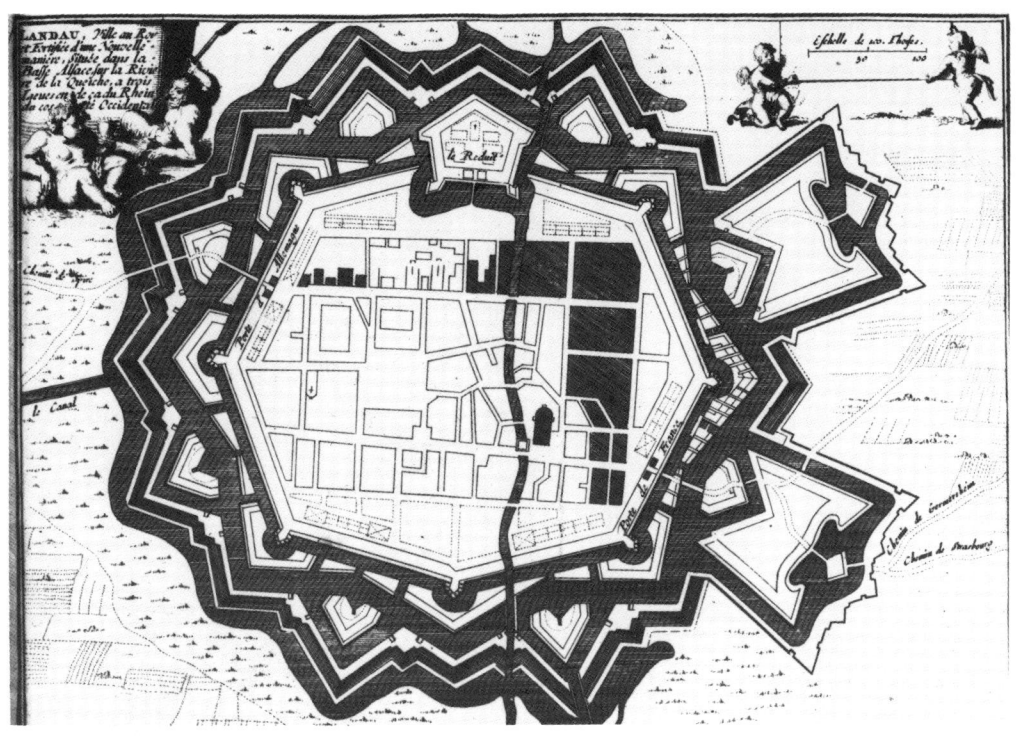

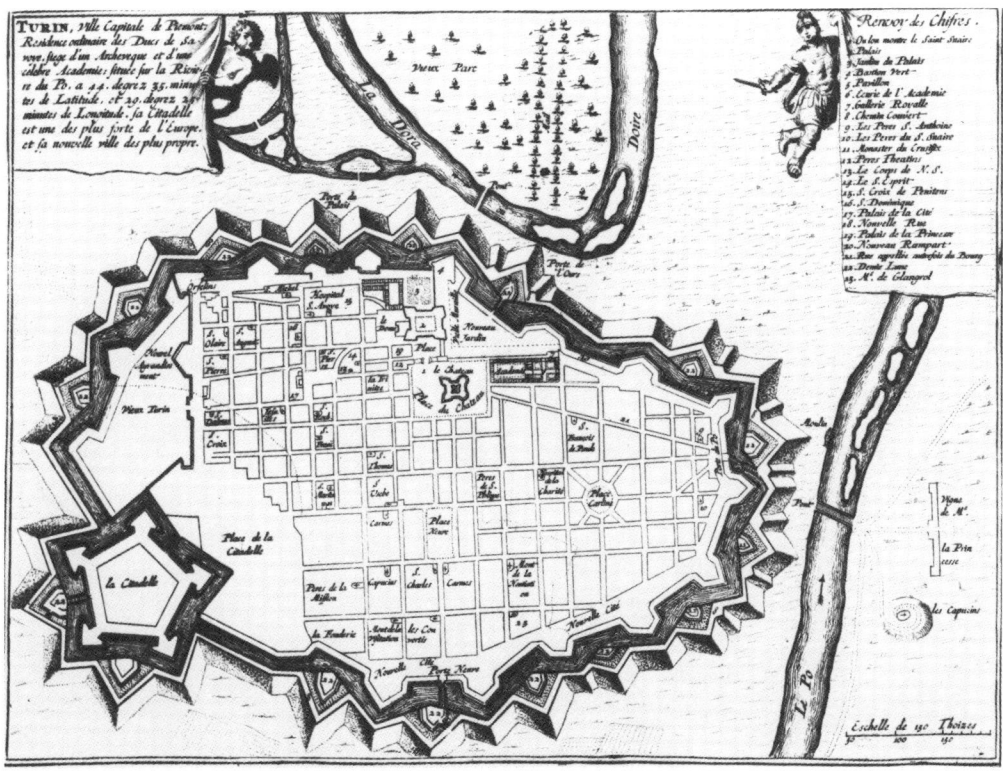

43

22 Floriani, Pietro Paolo, 1585–1638.
Diffesa et offesa delle piazze.
Macerata, Giuliano Carboni, 1630.

21.9 × 32.5 cm. Contemporary limp vellum, cuts and stain on lower cover, Camillo Giuseppe Corpignani 15 July 1669 autograph inside upper cover.
213 pages. Woodcut frontispiece, table of contents, subject index, 50 numbered woodcuts of materials and methods of siege and assault (7 illustrations in text, 43 full-page illustrations), illustration facing page 50 not numbered but counted as part of paginated text, 62 × 88 to 184 × 270 mm.

First Edition. Dedication to Ferdinand II, Holy Roman Emperor, is not dated, but printer's preface is dated 15 May 1630.

REF: Architekt und Ingenieur, Cockle, Miscellanea 6
COPIES: BN (Macerata 1630, frontispiece with author's portrait dated 1628; Venice 1654), HAB (Venice 1654), BM (1654), IU (1654), MH (1654), NN (1654), DLC (1654), MnU (1654), N (1654), WU (1654), MB (1654), MiU (1654).

FLORIANI IS AN IMPORTANT REPRESENTAtive of the seventeenth-century Italian fortification school, and his treatise became one of the few standard Italian texts on fortification of his time. Educated by the Marchese Alessandro Pallavicino and Giovanni de' Medici, he passed into the imperial service of Ferdinand II in 1619. Following the example of his father Pompeo Floriani, a colonel in the army of Rudolf II, Floriani served on assignments in Germany, Bohemia and Hungary, and was consulted about the fortifications of Vienna. During his Spanish service in 1617 he assisted at the siege of Vercelli under Don Pietro of Toledo. Floriani restored Vercelli following the Piedmontese surrender after 64 days of siege. In 1618 he was sent by the Spanish government to investigate the fortifications of Algiers, and in 1623 he was involved in the religious wars of Valtellina. A 1627 patent letter by Taddeo Barberini, chief commander of the papal army, named him vice-governor of Castel Sant'Angelo in Rome; in 1629 he was made commander of the papal armies in Umbria. In 1634 he fortified Ferrara, and in 1635 began the fortification of Valletta at the request of the Order of the Knights of Malta. His works there achieved great fame and are known in the annals of fortification as *Floriane;* they brought him great notoriety, even though the publication of his treatise was initially opposed by Pope Urban VIII for strategic reasons.

Displaying his knowledge of ancient authors, Floriani quotes generously from Greek and Latin authorities, but focuses closely on the defense and attack of fortresses. Among the topics he considers are the treatment of the surroundings of the fortress (advising destruction of unnecessary buildings), the quality and quantity of soldiery, and the handling of suspect citizens. The illustrations for the treatise include plans of Innsbruck, Altenburg and San Germano in Piedmont, and thus represent his own experiences in military service.

23   **Fournier, Georges,** 1595/99–1652.
Traité des fortifications ou architecture militaire tiré des places les plus estimées de ce temps, pour leurs fortifications, divisé en deux parties.
Paris, Iean Henault, 1661.

7.4 × 11.1 cm. Modern red calf.
ii + 186 pages + vi. Engraved title-page, preface, 110 numbered engraved full-page illustrations, 42 plans of fortresses, 3 tables of calculations, 65 fortification and bastion studies in plan and profile, 51 × 91 to 51 × 97 mm.

Third Edition. Dedication to François de l'Aubespine, Marquis de Hauterive, colonel of the French Army in Holland and Governor of Breda is dated 14 May 1648; the royal license to publish is dated 16 September 1647, the date of the first edition is 7 November 1647.

REF: Dainville, Marini
COPIES: MiU (Amsterdam 1668), N (1654), OrU (1654), MH (1654), IU (1654), PPAmP (1654), ViW (Leipzig 1688), PPL (Paris 1697).

SON OF A PROFESSOR OF LAW AT THE University of Caen, Fournier studied philosophy with the Jesuits at La Flèche, and then at Tournai, Bourges, Lille and Rouen. Between 1629 and 1635 he taught mathematics at La Flèche; in 1635 he was chaplain in the royal navy based at Dieppe; from 1640 to 1642 he taught mathematics at Hesdin, between 1645–46 he was prefect of the college at Caen, then at the college of Orleans in 1648–49. His publications range from the exhaustive *Hydrographie contenant la théorie et la pratique de toutes les parties de la navigation* (Paris 1642) to studies in coastal geography (*Geographia orbis notitia per littora maris,* 1648) and geometry (*Euclidis sex primi elementorum geometricorum libri,* 1644, French edition 1654, and three English editions).

The treatise would seem very useful not least because of its pocketbook format, and timely because of the fortification plans of Dutch towns it contains. His military art is composed of five parts: how to build and fortify all manner of sites, how to choose, train and lodge the soldiers, how to hold onto a place in peace as well as war, how to lay siege, and how to use firearms and fireworks.

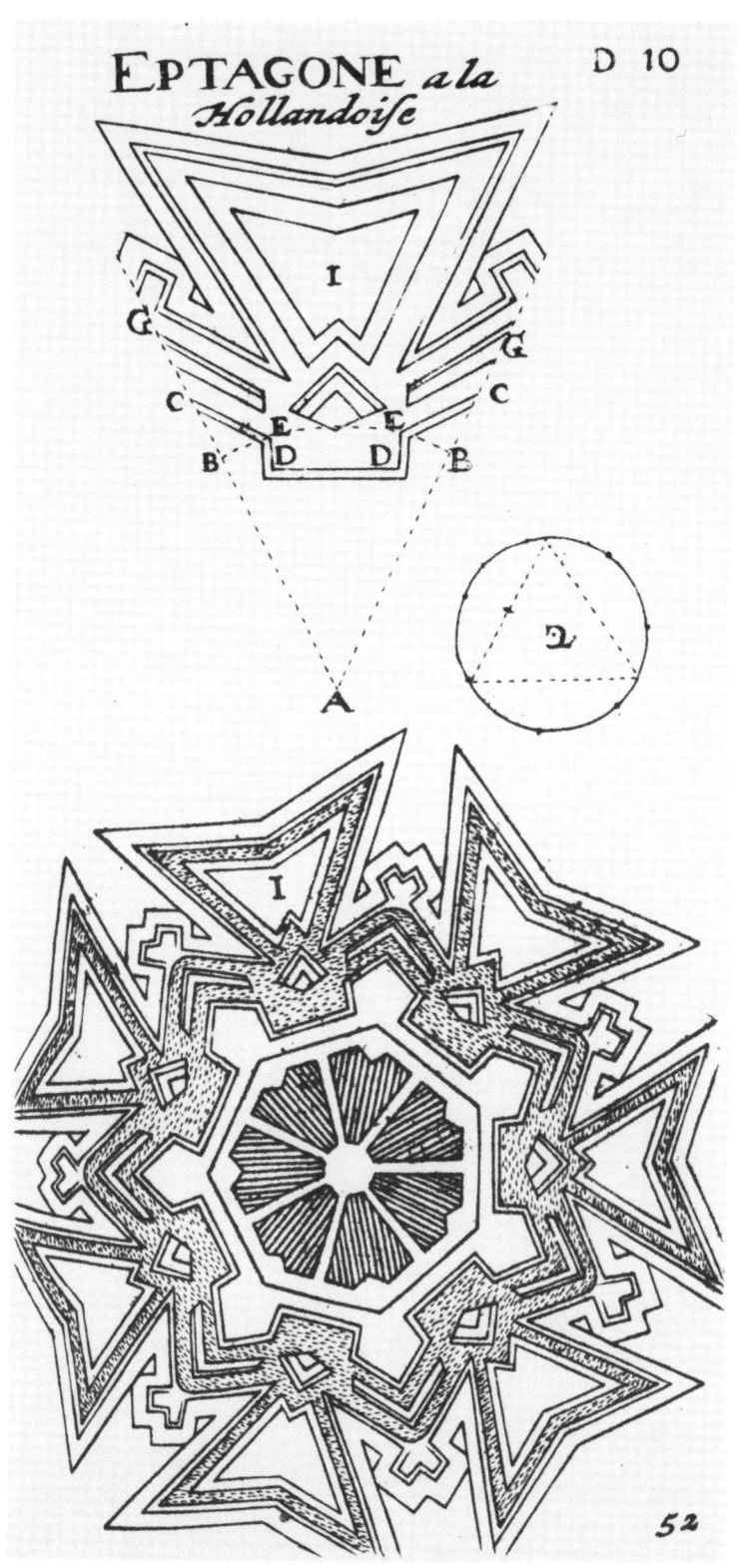

24 Freitag, Adam, 1602–1664.
Architectura militaris nova et aucta, oder newe vermehrte Fortification.
Leiden, Bonaventura and Abraham Elzevier, 1642.

19 × 28.2 cm. Brown half calf, tan and black speckled paper boards; title, author, date and "Kriegs Bibliothek" goldstamped on spine; K und K Kriegs-Archiv shelf-list label, ex-libris de Schallenberg, Louvain 1670 and "K. und K. Kriegsarchiv" stamped on title-page.

viii + 194 pages. Elaborate engraved title page, table of contents, 7 double-pages of calculation tables, 185 numbered engraved figures, plans and sections of fortresses and encampments on 30 double-page sheets, 309 × 235 to 327 × 243 mm, 8 pages of tables.

Third Edition. Dedication to Ladislaus Sigismund, King of Poland, is dated 1 July 1630.

REF: Architekt und Ingenieur, Cockle, Manno
COPIES: HAB (Leiden 1631, 1635 [French and German editions], 1642; Paris 1640; Amsterdam 1665), BM (1631, 1642, 1665; French editions, Leiden 1635 and Paris 1668), BN (French edition, 1635), DLC (1631), PU (French, Leiden 1635; 1642; Amsterdam 1665), MnU, DFo (Amsterdam 1665), MiU (Amsterdam 1665), CtY (French, Leiden 1635), NN (French, Leiden 1635), MH (French, Leiden 1635).

Freitag (or Fritach) was a physician, a philosopher and a mathematician. He spent time in the Netherlands where he assisted at the sieges of s'Hertogenbosch in 1629 and Maastricht in 1632. The title-page of his treatise represents Mars, the god of war, flanked by the muses of Geometry and Architecture, the two constituent elements of military fortification.

This lavishly illustrated treatise is considered one of the most influential published in the seventeenth century. Freitag's treatise definitively interpreted the Dutch fortification style and made it available to a large German public. It is divided into three parts focused on regular fortification, irregular fortification and the layout of encampment in the field. The first part includes a concordance of fortification nomenclature in French, German, Dutch and Latin. Freitag favors the Dutch manner of fortification construction, which holds that the ramparts should be covered with earth rather than clad in brick, and that moats should be wet or very wide and deep. In his plans there is a strict geometrical correspondence between gates, curtain walls, bastions and the orthogonal layout of the town.

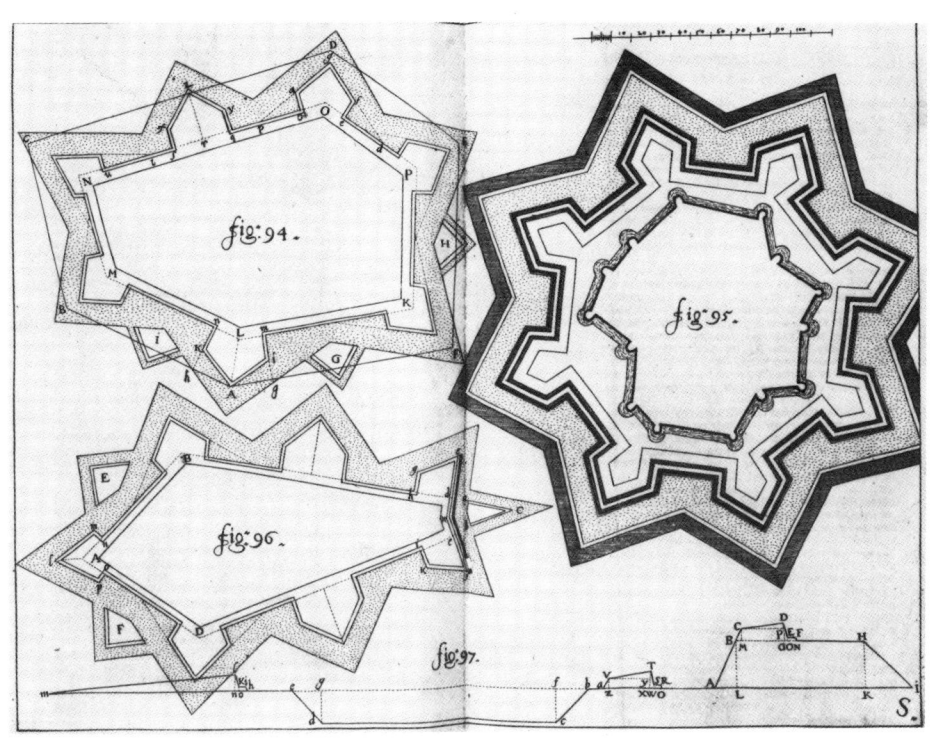

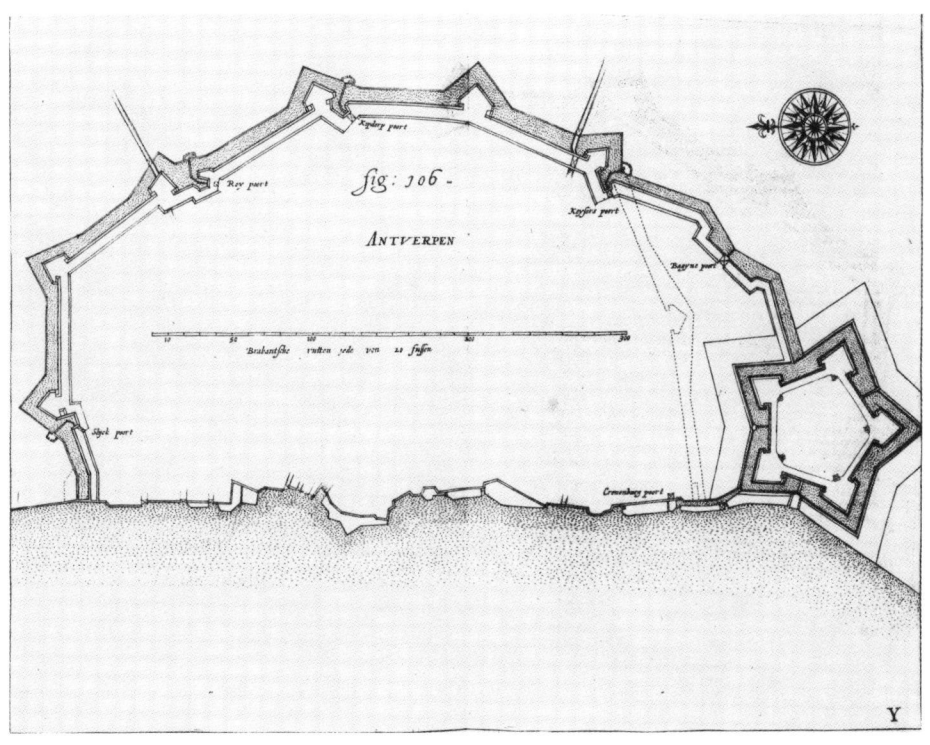

25 **Freitag, Adam,** 1602–1664.
L'Architecture militaire ou la fortification nouvelle.
Paris, Toussainct Quinet, 1640.

20.4 × 30.8 cm. Limp vellum.

viii + 179 pages. Engraved title-page, table of contents, 8 double-pages, illustrating fortress layout and section, siege materials, encampment plans, 66 × 89 to 298 × 238 mm.

Second French edition. Dedicated to Monsieur de Monceaux, royal councillor, by the publisher.

COPIES: N, CtY, NBU (Paris 1668), NN (Paris 1668), MH (Paris 1668).

The dedication is made to a court official even though his responsibilities were far from military ones ("bien qu'il traite de la discipline militaire, qui est une profession esloinée de la votre") because it could provide him with instruction.

26 **Gentilini, Eugenio.**
Instruttione de' bombardieri . . . et un discorso intorno alle fortezze.
Venice, Francesco de' Franceschi, 1592.

14.4 × 20.2 cm. Tan limp vellum.
viii + 126 pages. Table of contents, 3 part title-pages, 30 woodcut diagrams of cannons and fortifications, 106 × 24 to 283 × 202 mm.

First Edition. Dedication to Almoro Thiepolo, *proveditor* of the Venetian republic, is dated 5 May 1592.

REF: Cockle
COPIES: BM (1592, 1598), NN

THIS WORK IS DIVIDED INTO THREE PARTS: the examination given by the Schiavina, the head of Venetian gunnery, to would-be gunners, an analysis of the examination by Gentilini, and a discourse about fortification written in a form of conversation between Gentilini and his brother, a captain and engineer in the service of Venice.

27 **Goldman, Nicolas,** 1611–1665.
La nouvelle Fortification.
Leiden, Elsevier, 1645.

20.2 × 31.6 cm. Limp vellum, "di Lelio Onetti 1655" on fly-leaf.
xvi + 224 pages. Title-page engraved by Adrian Matham, preface, table of contents, 205 numbered figures grouped on 92 woodcuts, in text line-drawings (some of the plates are repeated [Fig. B is reproduced 4 times]) of fortification plans and sections, drawbridges and gates, 181 × 136 mm, numerous tables of calculations.

First French Edition. First edition, in Latin, was published by the Elsevier press in Leiden in 1643. Dedicated to Frederic Henry, Prince of Orange, Governor, Captain and Admiral of the United Provinces of Netherlands.

REFS: Manno, Wiebenson
COPIES: BN (1645, Latin 1643), BM (1643), NjP, PU, IEN, NNC, CtY, MH, NIC, MiU.

GOLDMAN WAS A MATHEMATICIAN, ARchitectural educator, and civil and military architect, particularly known for his description of the construction of the Ionic volute. He was the author of numerous successful works on architecture published between 1649 and 1656, as well as the posthumous work *Vollständige Anweisung zu der Civil-Bau-Kunst* (1696) and several eighteenth-century exegetical editions by his interpreter Leonhard Christoph Sturm (1669–1719).

In his erudite preface to this treatise,

Goldman established the relationship between human history, city-planning and military architecture. The treatise is divided into four parts. Part one is about drawings and parts of the fortress, part two about profiles and the ichnography of the city, part three about stereometry and sciagraphy [p. 108: "la sciagraphie c'est la peinture d'un ouvrage, comme il se void quand il est achevé"], and part four about mechanical problems and the attack of fortresses. The author is fully aware of the irony implicit in his subject matter, recognizing the circularity of the argument within the composition of his own treatise: "nous avons achevé la façon de bastir les forts et les forteresses . . . après nous avons un peu monstré . . . la manière de ruiner les ouvrages des ennemis" (p. 224). He establishes the aims of fortification as analogous to architecture (p. 44: "nous avouërons que la mire de l'architecture, et la fin proposée, consiste en deux choses, a savoir en la force, et en la beauté des ouvrages; le mesme s'entendra de la fortification"), discusses the representational aspects of military architecture (p. 195: "l'invention de telles sciagraphies seroit bien propre, pour parer l'entredeux des fenestres en quelque gallerie d'un grand prince; chacun avouëra, sans doute, qu'on ne sçauroit y trouver une peinture plus magnifique, que la perspective des places gaignées sur l'ennemy et cela principalement en la cour d'un prince victorieux et triomphant") and designates leaders of states as his preferred audience (p. 198: "aussi avons nous fait ces discours, non comme pour enseigner les mathématiques; mais pour ceux qui ont les charges d'une république; pour leur servir à augmenter leur gloire").

28 **Groote, Alexander de, Baron.**
Neovallia dialogo, nel quale con nuova forma di fortificare Piazze s'esclude il modo del fare fortezze alla regale, come quelle che sono di poco contrasto.
Munich, the widow Anna Berghin, 1617.

22 × 32.7 cm. Contemporary limp, blindstamped vellum, repaired and reinforced at edges, stippled black and red page edges, "Pietro Carolo Vinscher" in ink on title-page.

x + 285 pages + i. Engraved title-page by P. D. I., introduction, congratulatory poems by Antonio Francesco Tacchini and Eugenio Visdomini of the Roco Academy in Parma, 52 engraved illustrations of fortifications and sieges, 182 × 166 to 425 × 321 mm, errata sheet.

First Edition. Dedication to Maximilian, Duke of Bavaria, is dated 9 September 1617.

REF: Cockle
COPIES: BN, MiU, NN.

A MINOR GERMAN ARISTOCRAT, DE Groote was a councillor of Maximilian, general captain of the duke's artillery and governor of the fortress of Ketzting. His treatise, modelled on the *galateo,* takes the form of a conversation, a pastime between courtiers awaiting the duke in his garden; he sends word that he will not be coming out because he is busy discussing certain fortification projects. 18 points for and against heavy fortification are made in the first day of debate and discussed in 18 subsequent days. Voiteo is against *fortezza regale* because it is too expensive, and too long to build, it taxes too heavily the subjects of the prince, and it increases the price of materials. Polemico attempts to prove the linguistic importance of the *fortezza regale,* which implies royalty and truth. Although randomly labelled and unevenly scaled, the illustrations in this "Discourse on the new wall" are distinguished by their aggressive bellicose expressiveness; they and Groote's suggestion that the bastions be replaced by ravelins are not diminished by Tensini's accusation of plagiarism.

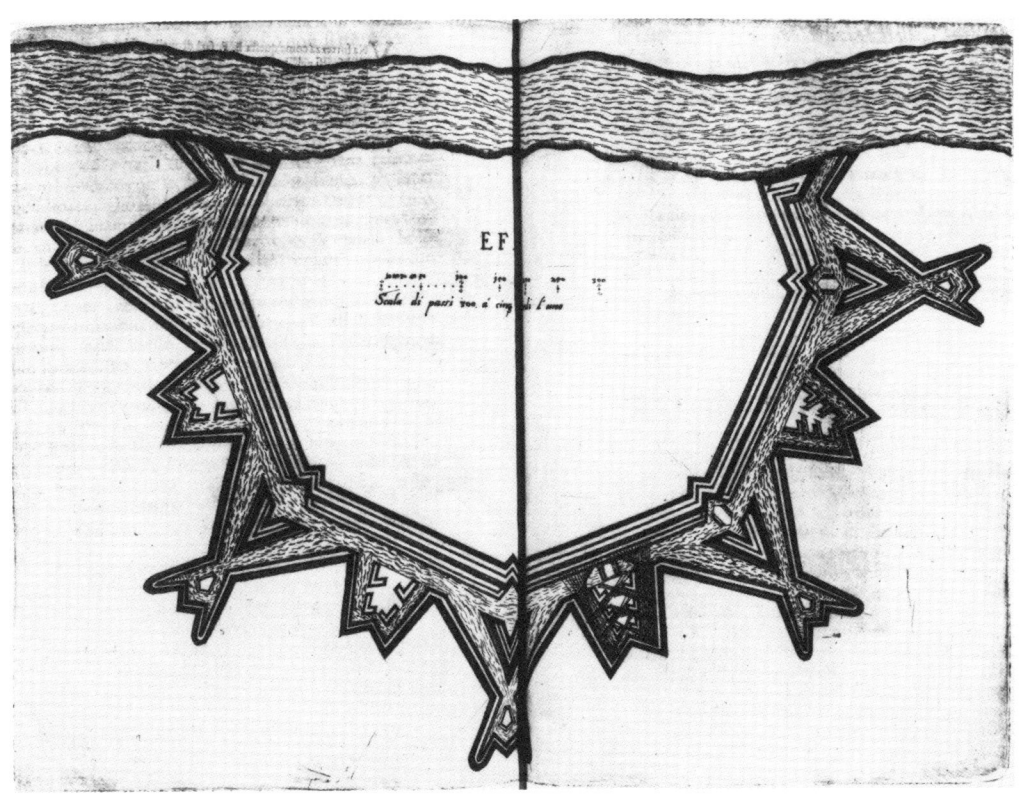

29 **Hondius, Henry,** 1573–1650.
Description et breve declaration des regles generales de la fortification.
The Hague, Henrik Hondius, 1625.

24.3 × 34.7 cm. Contemporary limp vellum, four leather ties (one removed), bound with 12-page *Breve instruction des regles de la geometrie, fort utile et necessaire a la perspective et architecture militaire, ou fortification,* Henry Hondius, 1625, with two plates in text.
viii + 108 pages + ii. Engraved title-page and part title-pages by Henrik Hondius ("Hh inventor" top right), preface, 24 engraved illustrations, 11 numbered illustrations in Part one, 4 illustrations in Part three, 9 illustrations in Part four (one seriously damaged), 105 × 98 mm to 882 × 300 mm.

First French edition. First edition in Dutch published by Hondius in 1624. Dedication to Christian IV King of Denmark and Norway is dated 15 May 1625 and proves that the brief treatise on geometry was intended to be bound together with the treatise on fortification.

REF: Architekt und Ingenieur, Cockle, Bagrow, Wiebenson
COPIES: BN, HAB (Dutch edition 1624, 1625), BM, NN, MH, DFo.

Hondius (DE HONDT) WAS ONE OF THE heirs of Jodocus Hondius (d. 1612), a publisher of maps in Amsterdam. After his father's death he continued the business in association with his brother-in-law Jan Jansson, publishing important maps and globes, including the Mercator-Hondius atlas.

Hondius's work is considered among the basic texts about the Dutch art of fortification, and like the treatises by Cellarius, Freitag, and Marolois it deals with every aspect of fortification science. The structure and content of the book, whose four parts are separated by part title-pages, are described in the preface. Part one focuses on the military architecture of regular and irregular places—from the triangular to the octagonal—and the transformation of a castle into a city. Part two is a treatise on artillery, describing different kinds of cannon. Part three considers the movements of the baggage trains and other matters of provision, while part four examines the layout of camps and deals with the attack and defense of fortresses, describing actual sieges (Julliers, 1610, Bergen-op-Zoom, 1622). Part four concludes the treatise with a study of fireworks.

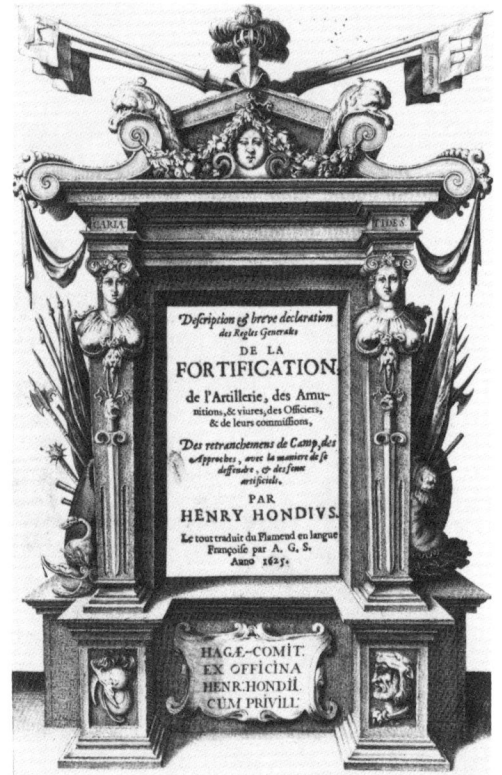

Hondius enters the discourse of fortification as though writing for a knowledgeable audience, or one aware of specialized terminology. He considers the work of Speckle, Errard, and Marolois excellent but far too expensive. For the design of gates he refers to Serlio, who had been translated into French by the painter Pieter Coeck d'Aelst of Antwerp. His writing ranges from general matters of method and fortification philosophy to detailed description of building materials. Thus we are informed that the citadel of Antwerp was built of brick while its walls were of white stone, but he also suggests that a prince should ask the opinion of many architects before he settles on a design in order to avoid errors (p. 15: "il seroit à desirer que quand un prince veut faire une forteresse, qu'il ne la fasse selon l'opinion d'un ou de deux, mais qu'il face faire autant de desseins qu'il trouvera d'architectes, pour prevenir les fautes qui arrivent le plus souvent"); his position on the location of the citadel—at the edge of the city—reinforces the established Renaissance opinion (p. 14: "touchant la maniere d'edifier un chasteau en une ville, chacun sçait qu'il ne doit estre au milieu d'icelle, pource que nul prince se pourroit mettre en sureté en iceluy, ny y faire entrer les vivres sans la permission de la ville.")

The beautiful siege maps in this treatise and the accurate city maps are no doubt due to Hondius's professional involvement in cartography.

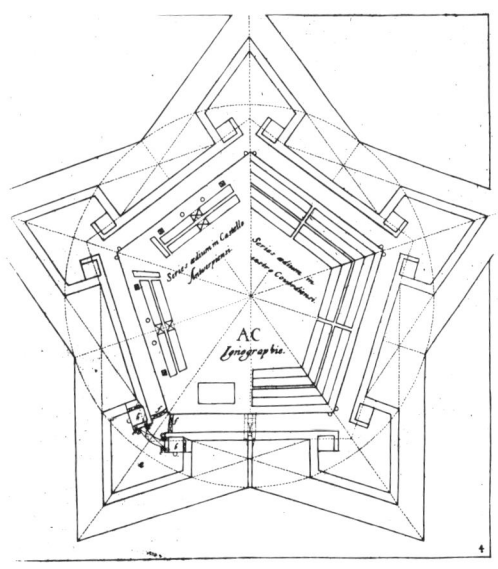
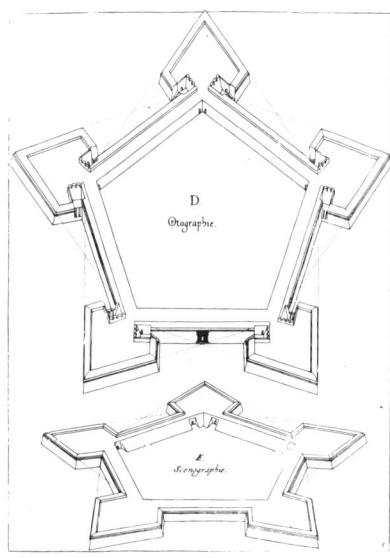

30 **Ive, Paul,** fl[ourished] 1602.
The Practise of Fortification: wherein is shewed the manner of fortifying in all sorts of scituations, with the considerations to be used in delining, and making of royal frontiers, skonces, and renforcing of ould walled townes.
London, Thomas Man and Toby Cooke, 1589.

13 × 18.5 cm. Vellum on paper boards, goldstamped title on brown panel.
iv + 40 pages. 12 full-page woodcut illustrations, 58 × 40 to 130 × 185 mm.
First Edition. Dedication to Sir William Brooke, Lord Cobham and to Sir Francis Walsingham, Privy Councillor.

REF: Cockle, DNB
COPIES: BM, MiU, DFo, NNC.

A MILITARY ENGINEER IN THE SERVICE OF the crown, Ive was a member of Corpus Christi College in Cambridge in 1560, but he does not seem to have matriculated. In 1597 he was paid for the fortification of Falmouth and for the transportation of prisoners to Spain, while in 1601 and 1602 he was employed in fortifying the isle of Haulbowline, near Cork. This copy is bound with *Instructions for the Warres* by Guillaume du Bellay [really by Raymond de Beccarie de Pavie, Baron de Fourquevaux], translated by Paul Ive (London 1589). Ive's treatise on fortification is divided into five chapters with the following titles: 1. The necessary placing of a fort. 2. The manner of fortifying in all sorts of grounds, and the commodities and discommodities a fort may have of it (sic) scituation. 3. The manner of the laying out of a fort, and the considerations to be used therein. 4. The manner of fortifying with earth. 5. The manner of fortifying of old walled townes.

31 **Lanteri, Iacomo de,** d. c.1560.
Due dialoghi del modo di disegnare le piante delle fortezze secondo Euclide; et del modo di comporre i modelli, et porre in disegno le piante delle città.
Venice, Vincenzo Valgrisi and Baldessar Costantini, 1557.

14.8 × 20.7 cm. Modern vellum, title goldstamped on brown leather label on spine.
viii + 95 pages + i. Preface, 33 woodcut figures, mostly explaining the geometrical basis of the polygonal bastion's design, 106 × 40 to 106 × 71 mm.
First edition. Dedication to Marcantonio Moro and Oliviero, Count d'Arco.

REF: Cockle, Vivenza, Miscellanea 12, Rocchi
COPIES: BM, Bodleian, NN (1601), MiU (1601).

LANTERI SERVED THE KING OF SPAIN AS engineer in Naples and on the coast of northern Africa where he drew up the plans of local fortifications. According to Promis, Lanteri also served Venice, the Papacy, and other princes.

This treatise was reprinted in 1559 in Venice and in 1583 in Rome, and again in 1601 in Venice. Lanteri is the author of another treatise, *Duo libri del modo di fare le fortificationi di terra intorno alle città*, published in 1559 in Venice by Bolognino Zaltieri, and then in a Latin translation in 1563 (*De modo subsistendi terrena munimenta atque oppida ceteraque loca omnia quibus aditus hosti praecluditur*)—dedicated to the Emperor Maximilian—by Valghisi in Venice (copy at Harvard), and *Della economica* (Venice, 1560).

Equally modest and sparsely illustrated, both studies are concerned with the geometrical foundation of Lanteri's proposed designs for the polygonal bastion. Lanteri is not at all concerned with illustrating the interior layout of the fortified city or fortress. His work was nonetheless very influential because he was one of the first writers to concern himself entirely with fortification and the polygonal bastion.

This treatise is composed as two separate conversations in which the participants are Girolamo Cataneo, Francesco Trevisi and a young man from Brescia (who is probably Lanteri himself, since his dedication to Count d'Arco is from Brescia). In the first dialogue military architecture is considered for the first time a branch of mathematical science rather than a form of practice, according to Promis, while Rocchi considers his fortification doctrinaire, based upon "sterile geometrical forces." But Lanteri's historical culture is evident in his criticism of curved bastions theorized by Dürer and Leonardo, and he regrets that the great poets Homer and Virgil never dealt with architecture, which he considers the most important human occupation after agriculture.

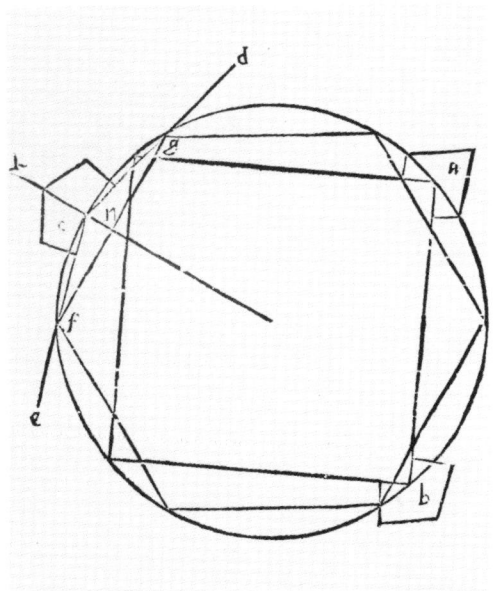

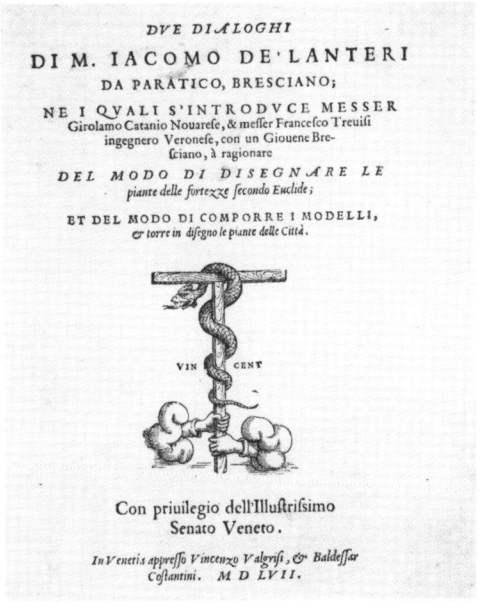

32 Lorini, Buonaiuto, c.1540–c.1611.
Delle fortficationi libri cinque.
Venice, Giovanni Antonio Rampazetto, 1596.

24.8 × 34.9 cm. Contemporary limp vellum, goldstamped decoration, stained lower cover, spine reinforced with mottled brown calf, title goldstamped on red morocco panel on spine, gilt edges.
xiv + 219 pages + 1. Title-page decorated with woodcut emblem, preface, table of contents, subject index, woodcut portrait of Buonaiuto Lorini (age 50), 115 woodcut illustrations, 142 × 130 to 503 × 353 mm.
First Edition, presentation copy. Dedication to Alfonso II d'Este, Duke of Ferrara, is dated 28 October 1596.

REF: Cockle, Manno, Architekt und Ingenieur, Miscellanea 14, de la Croix
COPIES: BM (1596, presentation copy to Vincenzo Gonzaga, same date; German edition, Frankfurt 1607), HAB (Frankfurt, German edition, 1607), Bodleian (1597), DLC (1597), NN, MiU (1597), WU (1597), OrU (1597), PPF (1597), MdBP (1597).

A FLORENTINE ARISTOCRAT, LORINI began his career in military architecture at the age of 22 in the service of Cosimo de' Medici, Grand Duke of Tuscany, who sponsored his studies in military engineering. After 1568 he spent four years with the Catholic army in Flanders where he assisted at the construction of Antwerp's citadel; in 1597 he was in Venice, and two years later he became an engineer of the Republic. His postings in the service of Venice included fortresses in Dalmatia, Zara, Legnano, Bergamo and Brescia, where he rendered the castle impregnable (except through treachery or starvation). He is considered a propagator of Savorgnano's theories; in 1592 they presented together the initial plans for Palmanuova, whose construction Lorini eventually supervised alone (from June 1600). According to Manno his autograph reports on Palma are preserved at ASVenezia, while his drawings of Palmanuova, Zara, Bergamo, Brescia, Crema and Orzinuovi were in the collection of Carlo Emanuele I, duke of Savoy. Lorini composed his treatise mostly in Zara and initially issued only presentation copies, but fearing counterfeit editions, he published it in 1597.

The treatise, divided into five books, is written in the form of a dialogue with a "count" whom Promis identifies as Nestore Martinengo, a well-known military personality. The author attempts to marry theory and practice in examining the fortification of cities and other sites. Book one investigates the principles of geometry, plans of bastions and design of fortresses; book two calculates the expenses of construction, and the design of bridges; book three is about various plans and moats; book four considers irregular fortresses; book five deals with mechanics, showing the acknowledged influence of Guidobaldo del Monte. The presentation of the arguments and of the scaled illustrations is done with thorough effortlessness, reinforced by the visual appearance of the page lay-out.

DELLE FORTIFICATIONI
DI BVONAIVTO LORINI,
NOBILE FIORENTINO.
Libri Cinque.

NE' QVALI SI MOSTRA CON LE PIV
facili regole la Scienza con la Pratica, di Fortificare le
Città, & altri luoghi sopra diuersi siti.

CON TVTTI GLI AVVERTIMENTI, CHE PER
intelligenza di tal materia possono occorrere.

Et il particolar soggetto di ciascun Libro si dimostra nel rouerscio di questa Carta.
NVOVAMENTE DATI IN LVCE.
Con Priuilegio.

IN VENETIA,
Appresso Gio. Antonio Rampazetto. MDXCVI.

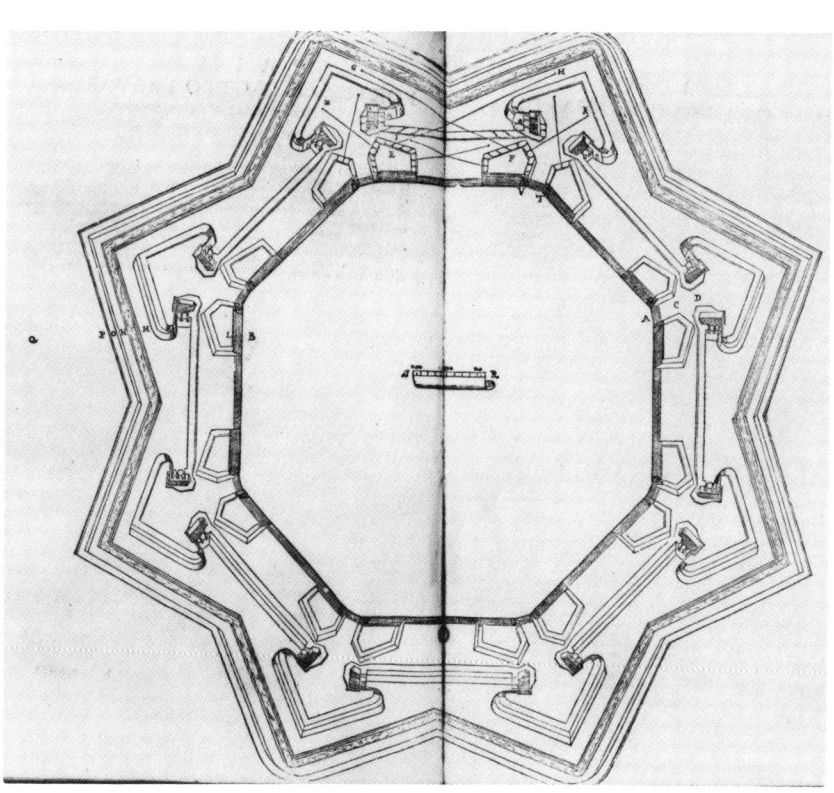

33 **Lorini, Buonaiuto,** c. 1540–c. 1611.
Le fortificationi nuovamente ristampate con l'aggiunta del sesto libro.
Venice, Francesco Rampazetto, 1609.

25.5 × 37.5 cm. Modern half calf, uncut edges.
xii + 303 pages (57–60 missing) + i. Title-page and part title-page for Book six, preface, table of contents, subject index, portrait of Lorini (age 60) engraved by W. Kilian, 150 woodcut illustrations, 122 × 118 to 483 × 371 mm.

Second Edition, corrected and enlarged. Dedication to Serenissimi Principi d'Italia is dated 25 February 1609. The sixth book is dedicated to Cosimo II de' Medici Grand Duke of Tuscany, also dated 25 February 1609.

COPIES: MCM, HAB (sixth book, Frankfurt 1616), Bodleian, Marciana, DLC, NcD, MB, PSt, OrU, MCM, CLU, MnU, WU, DNW.

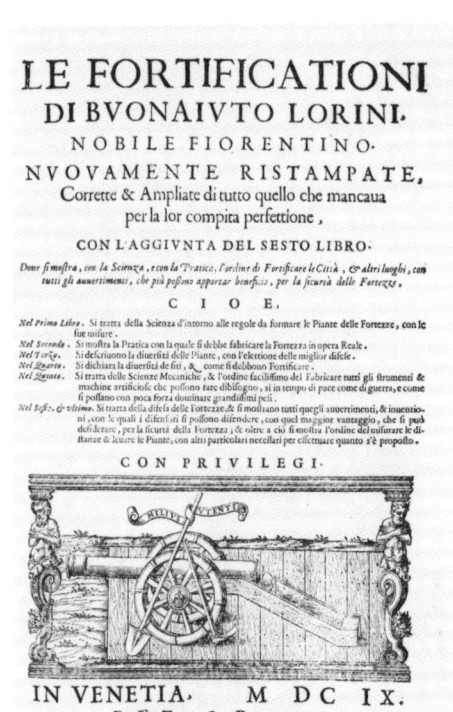
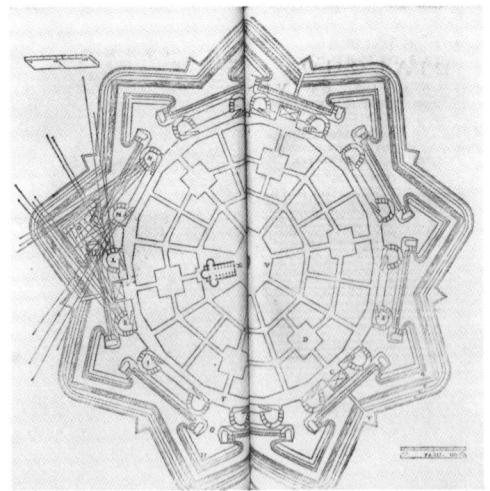

34 Lupicini, Antonio, c. 1530–c. 1598.
Discorsi militari sopra l'espugnazione di alcuni siti.
Florence, Bartolommeo Sermartelli, 1587.

14.7 × 21.8 cm. Off-white paper boards.
84 pages. Medici coat-of-arms and cardinal's hat on title-page.

First Edition. Dedication to Don Ferdinando Medici, Cardinal and Grand Duke of Tuscany, is dated 15 November 1587.

REF: Cockle, Miscellanea 14, Fortifications
COPIES: BM (1587, 1601), NN (1601), MiU (1587, 1601), MB (Venice 1840).

A FLORENTINE MILITARY ARCHITECT and hydraulic engineer, gunner, astronomer and mathematician, Lupicini participated in the war of Siena in 1552 at the sieges of Montalcino and Monticchiello. In 1578 he was sent to Rudolph II by his employer the Grand Duke of Tuscany; in 1584 he was in Venetian service having been requested by the Senate. In 1589 he worked on the channeling of the Arno in Florence and the bonification of the Maremma marshes. In 1594 he accompanied Giovanni and Antonio de' Medici to Hungary, acting as the latter's tutor.

Lupicini is also the author of the *Architettura militare con altri avvertimenti appartenenti alla guerra,* published by Giorgio Marescotti (Florence 1582), which must be the earlier book on the defense of fortresses dedicated to Francesco de' Medici, Grand Duke of Tuscany, that Lupicini mentions in his dedication of the *Discorsi* to the Cardinal, which was the second part of *Architettura militare*. It was subsequently republished in Venice 1601 together with Zanchi's *Modo di misurare le piante* and Lanteri's *Modo di fare le fortificationi.* His other publications bear witness to his numerous works as a hydraulic engineer.

In this small publication Lupicini deals with the siege of 30 fortresses; in 30 chapters the discussion follows a trajectory of increasingly difficult sieges, from the weakest to the strongest fortresses. Among his examples are the fortress in the plain with old-fashioned enclosure ("fiancato all'antica"), mixed fortification (old and modernized), the new bastioned system, cities by the sea, surrounded by marshes, on an island, bisected by a river, by a lake, with wet and dry moats, or on a hill, a port or an estuary. He also considers long-term siege by a large army waiting to starve the defenders of the fortress.

35 Maggi, Girolamo, d. 1572 et al.
Della fortificatione delle città libri III.
Venice, Camillo Borgominiero, 1583.

25.5 × 36.8 cm. Yellowed limp vellum, with cuts, gashes and torn corners.
iv + 136 leaves. Woodcut cartouche at top of title-page, table of contents, 32 illustrations of battle formation, 173 woodcut figures of which 6 are double-page, 37 full-page, 92 × 102 to 500 × 366 mm.

Second Edition. Dedication to Ferdinand, Archduke of Austria is dated 25 January 1583.

REF: HCL, Michaud, Miscellanea 1, Fortifications
COPIES: MH (1564), BM (1564), Bodleian, HAB, DLC (1564), NN (1564, 1583), MiU, CSt, CU (1564), DFo (1564), NjP (1564).

IN ADDITION TO MAGGI'S TREATISE, THIS work contains a discourse by Captain Francesco Montemellino on the fortification of the Borgo of Rome, dedicated to Duke Ottavio Farnese (1568), a treatise on infantry exercise by captain Giovacchino da Coniano, dedicated to Francesco della Torre, and an analysis of the fortresses of France by captain Jacopo Castriotto.

Maggi of Anghiari studied at the universities of Perugia, Pisa and Bologna, where he was a student of Robertello. While at Pisa he amused himself in the study of military architecture, antiquities and medals, the last inspired by Alciati's emblematic studies. He was appointed judge in Cyprus by the Republic of Venice and found himself besieged in Famagusta. He delayed the taking of that city through his ingenious mechanical inventions, but when the city finally fell to the Turks, Maggi was taken prisoner and sold to a ship-captain who took him to Constantinople. While in prison he composed *De tintinnabulis* (Hanau 1608, Amsterdam 1664) and *De equuleo* (Hanau 1609) dedicated respectively to the ambassadors of the Emperor and of the King of France who attempted to have him freed. He was discovered after his escape and strangled in a Turkish prison.

Maggi's other literary works include *I 5 primi canti della guerra di Fiandra,* edited by Pietro Aretino (Venice 1551), *De mundi exustione et de die judicii* (Basel 1562), *Variae lectiones seu Miscellanea* (Venice 1564) and many manuscript writings including "Degli ingegni e secreti militari."

Castriotto, born Jacopo Fusto, was the scion of a patrician family from Urbino. He studied military and civic architecture with Girolamo Genga, whom Lomazzo called "architetto universale." He served the dukes of Urbino and then the Spanish crown in Naples as captain and engineer, where he also married a daughter of Scanderbeg and thus added her name Castriotta to his own. In Rome from 1542 he was employed for eight years in fortifying the city; he designed simultaneously the fortification of Sermoneta, a town south of Rome, at the request of the Gaetani, and in 1548 he assisted in the debates among the best engineers of the day regarding the fortification of the Borgo, presided over by Pope Paul III. In charge of the siege of Mirandola (1551), he was present at the sieges of

Montalcino and Monticchiello (1553), and employed by the King of France from 1554. In St. Quentin in 1556, he left the city before the terrible defeat suffered by the French, and participated in numerous other sieges in Picardy until the peace of 1559. He achieved the rank of general in the French service and died in Calais. Castriotto claims to have made many drawings and models of fortresses for Henry II (p. 19). His correspondence on military subjects with Tartaglia is preserved in the Urbino archives; his other professional acquaintances included Alghisi and de' Marchi, both of whom he met in Rome during the papal-sponsored debates.

Maggi begins his treatise with a learned discussion of the origins of the city and its optimal size. His history of settlements and plans of early cities is based upon his reading of the ancients; among the authors he quotes are Pliny, Virgil, Aristotle, Cicero, Sallust, Plutarch, Varro, Homer, Vitruvius and Vegetius. His discourse is not arid, being rendered topical by continuous references to contemporary thinkers and builders of cities and fortifications. He mentions the famed architect Michele Sanmicheli, Lanteri, Cataneo, and Tartaglia, as well as important current or recent events such as the fortification of Rome in the 1540s, and the battle of St. Quentin. One of the reasons that this treatise is significant is the large cast of characters and the distinct historical framework within which it is composed.

Maggi has well-formed opinions about a number of key issues for fortification strategy and city planning. He contends that the citadel should be part of a fortified town, and that the for-

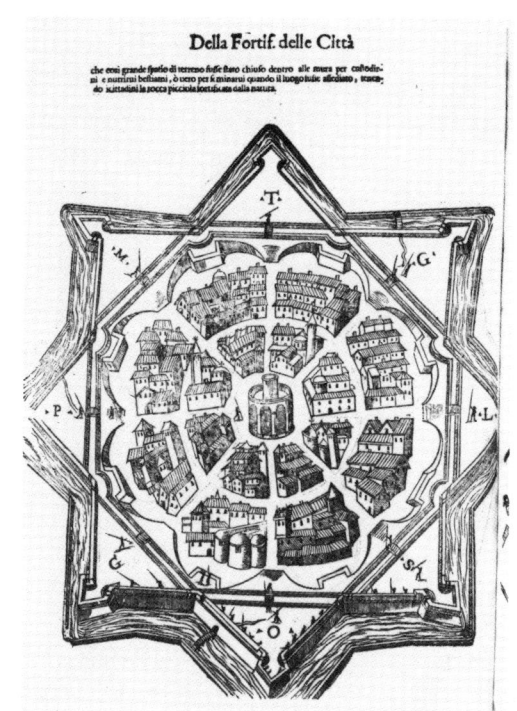

tification should rely on a single enclosure of walls; he advises against straight, wide streets—even though they make a city beautiful—because narrow streets keep the city cooler and indirect streets prevent the enemy from finding his way directly to the center of town. He believes the form of the fortress to be more important than the materials used to build it, and that cannon should be used directly against the enemy rather than as flanking fire. Both he and Castriotto suggest that the curtain wall between two bastions need not be straight or continuous, since the angled wall allows far more opportunities to install cannons, thus increasing the ferocity of the fortress's firing range. Their designs are generally over-structured and over-armed, with heavy masonry structures built to support double casemates and gun-emplacements; the results are heavy and menacing architecturally as well as militarily.

36 Malthus, Francis.

Traité des feux artificiels pour la guerre, et pour la recréation; avec plusieurs belles observations, abbrégez de géometrie, fortifications et examples d'arithmétique.
Paris, Pierre Guillemot, 1629.

11.2 × 17.4 cm. Red morocco, goldstamped decoration and title on spine, ex-libris DR, gilt edges. xii + 243 + xi. Engraved title-page, preface, 37 engraved illustrations, 52 × 52 to 107 × 173 mm, table of contents.

First Edition. Dedication to Cardinal Richelieu; royal license to publish dated 24 December 1628.

REF: Cockle
COPIES: BN, BM (1633), MH (1629, 1632), CtY (1629, 1632), CLU (1640), NcD (1640), MiU (1661).

FRANCIS MALTHUS, SIGNED AS SR. F.D.M. on the title-page, an Englishman, also published an English version of this treatise entitled *A Treatise of Artificial Fire-Works Both for Warres and Recreation: With Divers Pleasant Geometricall Observations, Fortifications, and Arithmetical Examples* (London, 1629) as well as *La pratique de la guerre* (Paris, 1650 and 1668). He was an engineer in the royal French army and a captain general of mines and sapping, and was responsible for the French use of the mortar from 1634; his work was more sophisticated than that of contemporary English specialists since he received his education in pyrotechnics on the continent.

This treatise is divided into five parts: a discussion on the manufacture and use of mortars, grenades, fire arrows and petards, the art of making fireworks (golden rain, starbursts, and sausages are among the forms he mentions), a treatise on practical geometry (how to measure heights, distances, changes in level of the terrain, and the use of the compass), an essay on fortifications (regular and irregular), and a treatise on arithmetic. The brief section on military architecture, illustrated with bastion studies and fortification plans, is inspired by the French manner (right angle between curtain wall and the flanks of the polygonal bastion).

37 **Manesson Mallet, Allain,** 1630–1706.
Les travaux de Mars ou l'art de la guerre.
Paris, Denys Thierry, 1685 (but printed 15 November 1684), 3 volumes, volumes 2 and 3 dated 1684.

12.6 × 20.5 cm. Brown calf, goldstamped spine with title and volume on separate red morocco panels, ex-libris Prince Borghese (Burghesii) inside upper cover of each volume.
Vol. 1: xxvi + 363 pages + v. Engraved frontispiece dated 1684, portrait of Louis XIV engraved by P. Giffart and dated 1683, preface, table of contents, portrait of Manesson Mallet engraved by P. Landry and dated 1683, 152 engraved and numbered illustrations of fortification studies and plans and views of actual cities, 98 × 144 mm, subject index.
Vol. 2: xiv + 341 pages + iii. Engraved frontispiece, preface, table of contents, 111 engraved and numbered illustrations of fortification systems proposed by the great 16th- and 17th-century military architects (# 109 missing), 93 × 144 mm, subject index.
Vol. 3: xii + 387 pages + viii. Engraved frontispiece, table of contents, 137 engraved and numbered illustrations, mostly of soldiers, officers and siege entrenchments, 93 × 140 mm, subject index.

Second Edition, enlarged. Dedication to Louis XIV; published with royal license dated 19 August 1684.

REF: Manno, Architekt und Ingenieur, Pastoureau, Michaud.
COPIES: BN, HAB (1685, The Hague, 1696), MiU (1685), NN (1671), DLC (1685, 1691; The Hague 1696), DFo, NIC, CtY, NjP, NcU, NNC (Amsterdam 1684), OU (Amsterdam 1684), MH (1685, Amsterdam 1684), NbU (The Hague 1696), CLU (The Hague 1696), ViW (The Hague 1696), MnU (The Hague 1696), ICJ (The Hague 1696).

IN 1684 MANESSON MALLET HELD THE title of mathematician of the royal grooms, but previously, as engineer and sergeant-major of artillery for Alfonso IV king of Portugal, he built numerous fortifications. His other publications include *Description de l'univers, contenant les differents systemes du monde, les cartes generales et particulières de la géographie ancienne et moderne, les plans et les profils des principales villes et des autres lieux plus considerables* (Paris 1683, Frankfurt 1685) in 5 volumes and *La géometrie pratique* (Paris, 1702) in 4 volumes.

Manesson's treatise is outstanding for the thoroughness of his approach, the extensive illustrative material and the pocket-book format which allowed the text to be used in the field. Furthermore, his critical analysis and comparisons of the fortification systems proposed by his predecessors in the sixteenth and

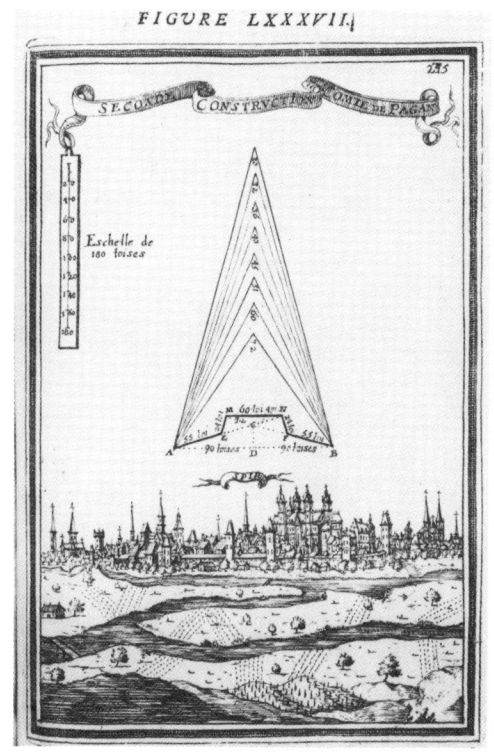

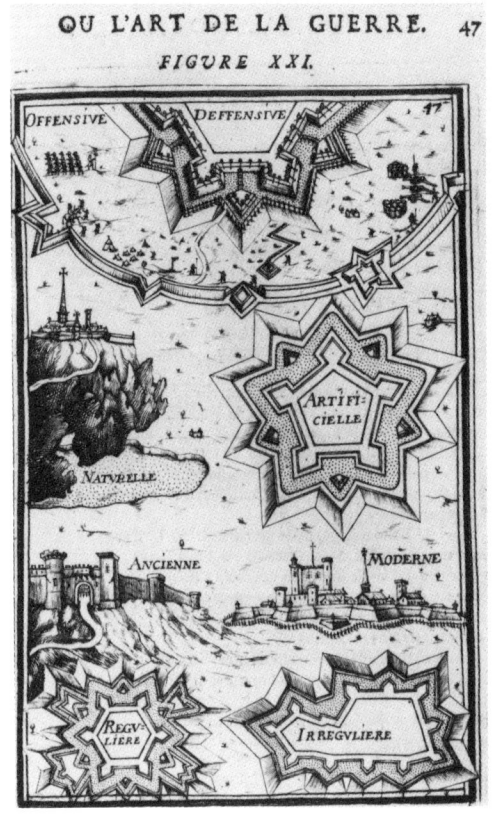

seventeenth centuries make his treatise into an invaluable historical resource as well. This work was a great success, prompting a counterfeit Dutch edition and many foreign language editions.

In the first volume he teaches how to fortify regular and irregular sites, in the second he compares the construction techniques of famous military architects such as Errard, Marolois, Freitag, Stevin, Dögen, de' Marchi, Sardi, Deville, and Pagan, while the third volume is dedicated to an analysis of the functions of the cavalry, infantry and artillery and outlines his method of attack and defense of a fortified site.

38 Manesson Mallet, Allain, 1630–1706.
Kriegsarbeit oder neuer Festungsbau.
Amsterdam, Jacob von Meurs, 1672.

11.3 × 18.2 cm. Modern half cloth, brown calf boards, K. und K. Kriegs-Archiv shelf-list label inside upper cover.
xvii + 130 pages + xiv + xiv + 195 — ix + xvi + 272. Dutch title-page engraved by R. de Hooghe, two part title-pages, preface, 3 part tables of contents and subject indices, 305 engraved illustrations (131 numbered in part I, 77 in part II with number of page adjacent to illustration, 97 in part III), 96 × 146 mm.

First German Edition. Dedication to Leopold, Holy Roman Emperor is dated 3 February 1672 in Vienna, and to Johann Kramprichen, Knight of the Order of S. Maurizio and S. Lazaro.

COPIES: HAB, NNC, TxC, CtY, NjP, CU, MH.

T HE ENGRAVINGS IN THIS EDITION ARE lighter in tone and are of better quality than in the French edition of 1685.

39 **Marolois, Samuel,** 1572–1627.
Fortification ou architecture militaire tant offensive que defensive.
The Hague, Henrik Hondius, 1615.

37.7 × 27.5 cm. Limp vellum, title goldstamped on spine, ex-libris Ducal Library of Brunswick inside upper cover.
49 leaves of text. Engraved title-page by S. Frysius 1615 cut and mounted, 164 engraved figures mostly of bastion studies and fortification plans on 40 numbered plates (numerous brown and violet stains), 324 × 232 mm.

First edition. This is the fourth part of the *Opera mathematica ou oeuvres mathematiques traictans de géometrie, perspective, architecture et fortification* published by Hondius between 1614 and 1616.

REF: Architekt und Ingenieur, Manno, Cockle
COPIES: BN, BM (1628, 1651), MiU, DFo, CtY, DN.

M AROLOIS WAS A FRENCH MATHEMATIcian who spent most of his active career in the Netherlands, where he cultivated strong ties with Vredeman de Vries and Henry Hondius. Marolois's fortification system was inspired by Francesco de' Marchi's work, which had been adopted by Dutch engineers. This treatise was a great success, as seen from the numerous editions and translations. In Amsterdam the treatise was published in Dutch, German and Latin in 1627–28 by Janssen; two Latin editions were printed by him in 1633 and 1644, and an English edition in 1638.

The beautiful title page is followed by a very technical work divided into two parts, on regular and irregular fortification. The text is a mere explanation of the figures; there are few theoretical precepts and only a few definitions of fortification elements. Thus the treatise seems clearly intended for a mature professional military architect and reflects a discourse internal to the discipline.

40 **Marolois, Samuel,** 1572–1627.
Fortification: wie ein Ort nach der wahren und fundamental Kunst zubefestigen.
Amsterdam, Jan Janssen, 1627.

19.2 × 29.4 cm. Reused parchment from a philosophical manuscript of the 13th-14th century, worn spine.
ii + 111 pages + v. Engraved title-page by W. Ackersloot, table of calculations, 164 engraved figures mostly of bastion studies and fortification plans on 40 double-page plates, 317 × 227 mm.

First German edition.

COPIES: HAB (1637, 1638), N, CtY, NN, MH, NNC, IU

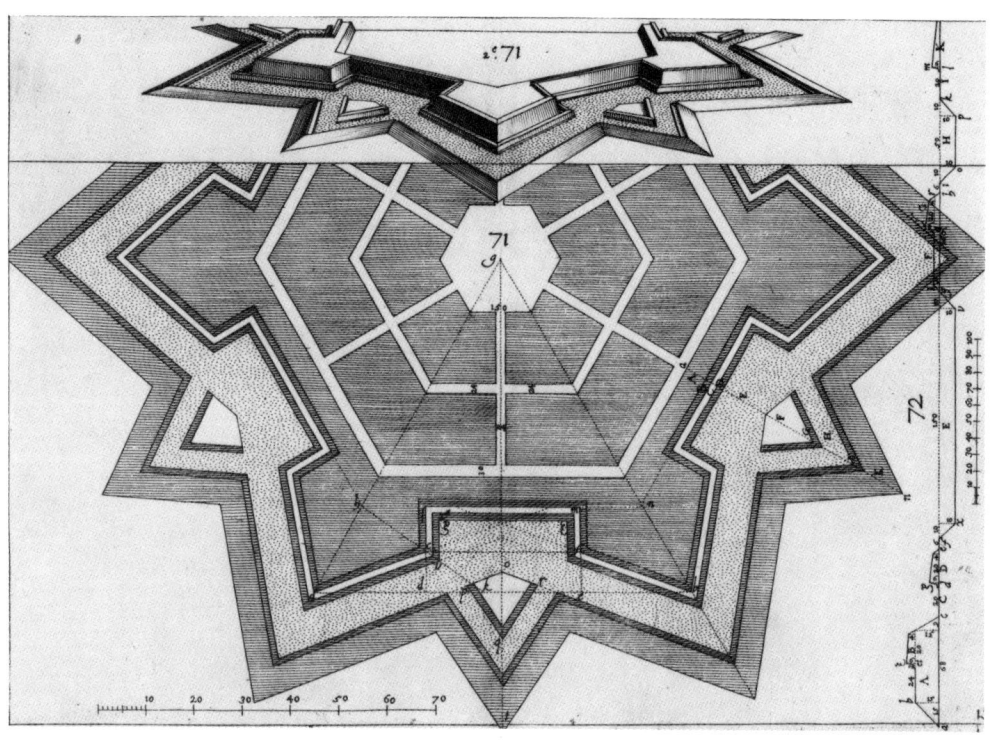

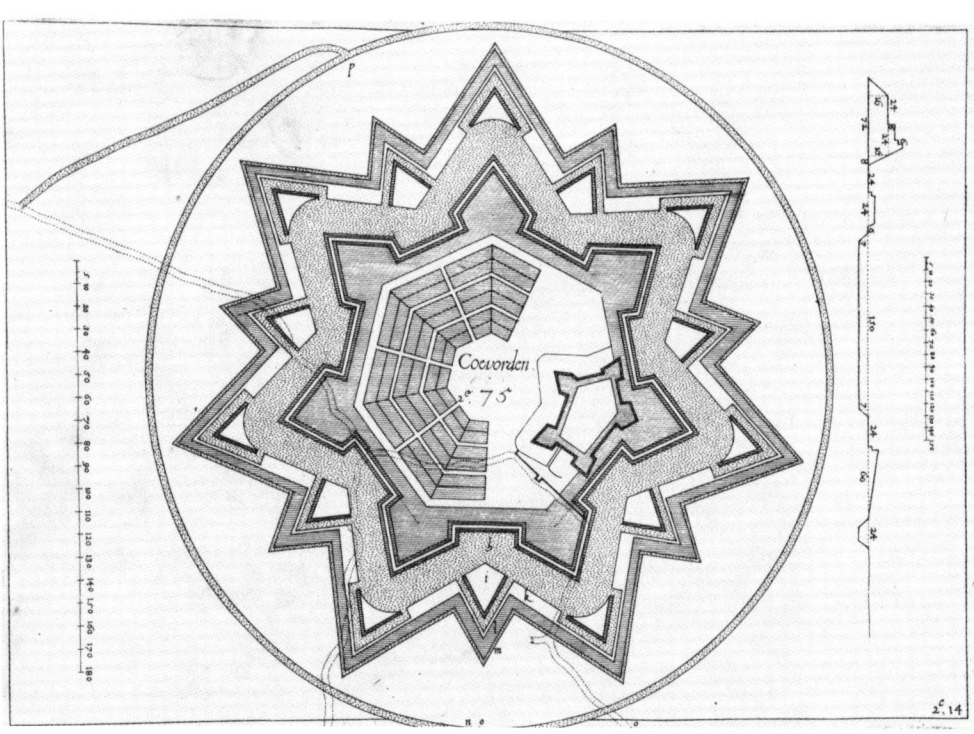

41 **Moore, Sir Jonas,** 1617–1679.
Modern Fortification, or Elements of Military Fortification.
London, Obadiah Blagrave, 1689.

1.5 × 17.2 cm. Brown calf, goldstamped decoration on spine and title on red morocco panel, ex-libris Cholmley Turner Esq. inside upper cover.
xvi + 127 pages + i. Table of contents, bibliography, 7 plates of figures, 131 × 85 to 205 × 212 mm.
Second Edition. Dedication to Sir Thomas Chicheley, Master-General of the Ordnance and Armory and Great-Master of the Artillery is dated 27 March 1673, the year of the first edition.
REF: Early Military Books, DNB
COPIES: MiU, MnU (1673, 1689), DFo, CLU-C (1673).

Moore, born in Lancashire, was the tutor in mathematics of the duke of York in 1647. He was a surveyor in the work of draining the Great Level of the Fens after 1659; Pepys was said to possess a copy of his survey charting the entire course of the Thames. Sent to Tangiers in 1663 to inspect its fortification, he was knighted upon his return and appointed Surveyor-General of the Royal Ordnance. He procured from the king the foundation of the Royal Observatory in 1674; the same year he entered the Royal Society. His splendid library of scientific works was housed with him in the Tower. His other publications include *A New System of Mathematics* (1681), *A Mathematical Compendium* (1674) and *A Treatise of Artillery* translated from the Italian work by Moretti but published only in 1683.

This work—"calculated for the meanest understandings"—is a very useful handbook, and sketchily covers a great number of subjects. What he says on p. 102—"Because I am tied to a short discourse I shall only give you the terms, and refer you to great volumes writ on this subject"—is applicable to every chapter; the illustrations are poor and their irregular numbering points to careless plagiarism, but he does offer a vivid comparison between the fortification theories and firearms strategies of numerous European masters.

42 **Mora, Domenico,** 1539–c. 1601.
Tre quesiti in dialogo sopra il fare batterie, fortificare una città et ordinar battaglie quadrate, con una disputa di precedenza tra l'arme e le lettere.
Venice, Giovanni Varisco e compagni, 1567.

15.1 × 20.5 cm. Modern grained beige calf.
iv + 68 leaves + ii. 4 woodcut illustrations of bastions, 106 × 35 to 108 × 54 mm, subject index.

First edition. Dedication to the Duke of Florence and Siena, founder of the order of Santo Stefano, is dated 9 March 1567. Secondary dedication to the members of the Academia dei Storditi in Bologna.
REF: Cockle, HCL, Miscellanea 4
COPIES: BN, BM, Bodleian, MiU, NNC.

MORA, AN ARISTOCRAT FROM BOLOGNA, was a member there of the Academia dei Storditi and was in the service of the Medici family in Florence. In 1570 he was in Venetian service at the garrison of Zante but was dismissed in disgrace. He participated in the religious war in Avignon and published an account of it. In 1579 Mora went to Poland to serve Stefan Bathory of Transylvania (elected King of Poland in 1575), on whose behalf he engaged in numerous sieges against the Russians. Devoted to his new homeland, he published a call for a crusade in Vilna in 1595. He authored *Il cavaliere in risposta del gentilhuomo* (Vilna 1589), in reply to Girolamo Mutio who in his *Il Gentilhuomo* (1565) had given precedence to letters, and earlier *Il soldato* (Venice 1570), a text on military art. He referred in 1589 to a work on the practice of war entitled *L'Imperatore,* which would have crowned his ranked publications. His writing is coarse, perhaps on purpose, since he seems to have been fairly well-read despite his disclaimers. This stylish roughness may have led Promis to place Mora among the writing soldiers rather than the cultivated and scholarly engineers, even though this work is more of an academic debate than an illustrated and practical manual.

43 **Mut, Don Vincente,** 1614–1687.
Arquitectura militar.
Majorca, Francisco Oliver, 1664.

13.8 × 20 cm. Modern black morocco, goldstamped title on spine.
iv + 158 pages. Table of contents, 44 woodcut figures on three folded plates, 294 × 190 mm.
First edition.
REF: EUI.

ALTHOUGH HE HAD JOINED THE JESUITS in 1629, Mut soon renounced the order. He studied mathematics and law as well as military art and became a sergeant-major in Palma and engineer of the kingdom of Majorca. He fought in the wars of Catalonia, and in his own time was considered an impartial historian, an erudite antiquarian, able theologian, excellent mathematician and inventor of mathematical instruments. His treatise focuses upon the fortification elements of regular and irregular fortresses and is divided into 39 chapters. His other publications include *El principe en la guerra y en la paz* (Madrid 1640), *Historia del reino de Mallorca* (Majorca 1650, Palma 1841) and *Adnotaciones sobre los compendios de la artilleria* (Majorca 1668).

44 **Ozanam, Jacques,** 1640–1717.
Traité de fortification contenant les méthodes anciennes et modernes pour la construction et la défense des places, et la manière de les attaquer, expliquée plus au long qu'elle n'a été jusques à present.
Paris, Jean Jombert, 1694.

12 × 17.7 cm. Limp vellum.
xxiv + 256 pages. Title-page engraved by Jan van Vianen, preface, table of contents, subject index, 44 full-page engraved diagrams of fortresses and bastions folded-in, 216 × 177 mm.

First Edition [pirated]. Dedication to Johann Adolph, Duke of Schleswig-Holstein and General of the United Provinces army, by the publisher Adrian Moetjens.

REF: Architekt und Ingenieur, Michaud
COPIES: HAB, BN, DLC, CLU-C (London 1741), CU (London 1741), MdBP (London 1741), MH (London 1741), MnU (London 1741).

OZANAM, BORN INTO A FAMILY OF Jewish origin, was an engineer and a mathematician, and a member of the Académie des Sciences. He published numerous works of theoretical mathematics, but also practical texts since he had to earn his living. He taught privately for the same purpose. Among his publications are the *Tables des sinus, tangentes et sécantes, et des logarithmes* (Lyons 1670, Paris 1685, 1720); *Traité de gnomonique* (Paris 1673 and 1685); *Géographie pratique* (Paris, 1684); *Traité des lignes des premier genre* (Paris 1687) and *Récréations mathématiques et physiques* (Paris 1694, 1720, 1735, 1741).

His treatise on fortification is divided into six parts. In the first part he deals with regular fortifications, preceding them with an explication of ichnographic fortification terms and the general maxims of fortification. Part two is about the foreworks; part three compares the different manners of fortification proposed by French (Errard, Pagan, Blondel, Vauban, Deville), Italian (Sardi) and Franco-Dutch (Marolois) military architects. Part four is about the fortification of irregular places, part five about offensive fortifications, and part six about defensive works. The value of this work consists in its summary of the major innovations in military architecture and siege strategy made in the seventeenth century.

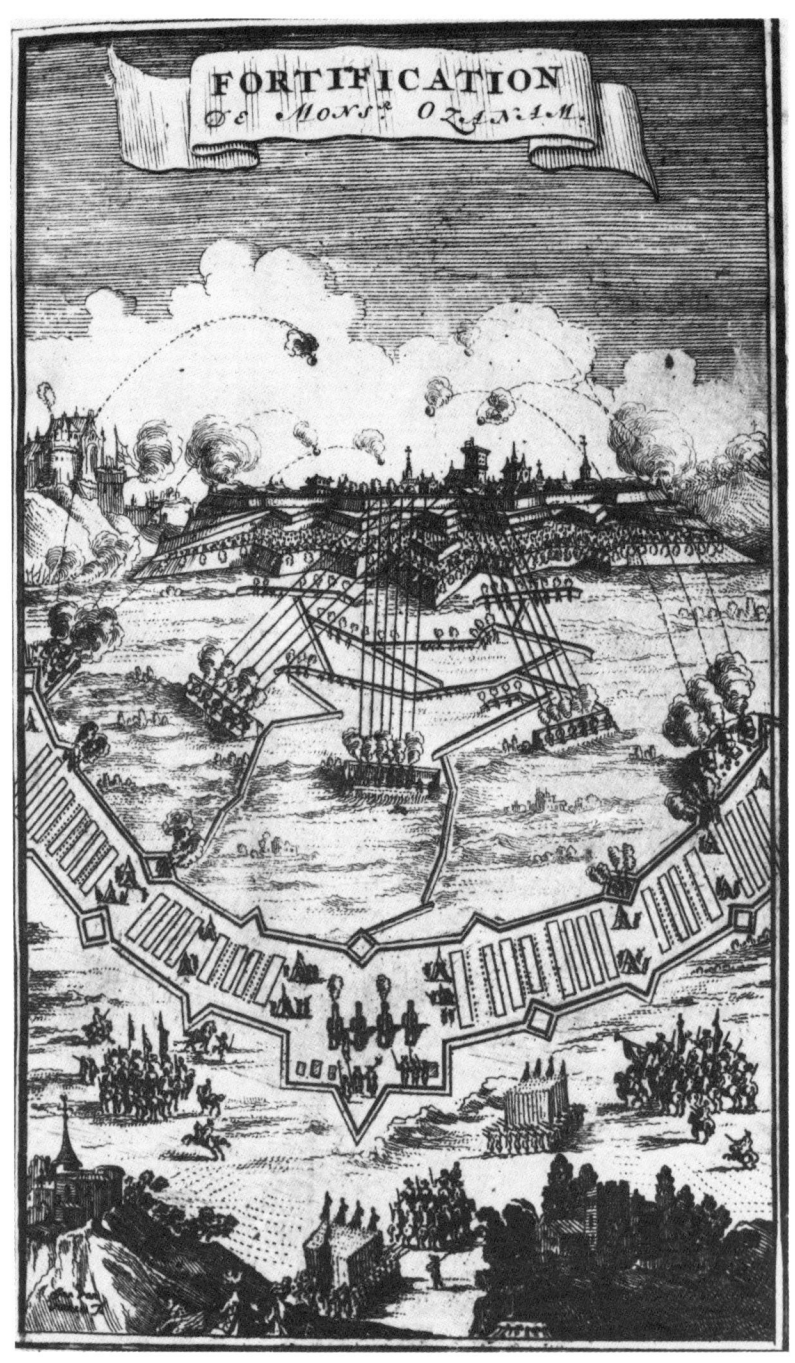

45 Pagan, Blaise François, Count of, 1604–1665.
Les Fortifications.
Paris, Cardin Besogne, 1645.

21.6 × 33.3 cm. Contemporary limp vellum, many wormholes, large engraved coat-of-arms of the author on title-page, marginalia in Italian.
iv + 116 pages + iv. Preface, 19 woodcut figures, 76 × 81 to 150 × 247 mm, two tables of calculations, table of contents.

First edition. Dedication to Hugues, Duke of Terranova, is dated 4 January 1645; the royal license to publish is dated 31 December 1644.

REF: Michaud
COPIES: BN (1645, 1668, 1669, 1689), Bibliothèque du Génie-Paris, MH, NWM, MiU (Brussels 1668), NjP (Brussels 1668).

An engineer and astronomer, Pagan was born into a noble family from Avignon of Italian descent—as his extensive genealogy at the beginning of the work demonstrates (Hugues Pagan or de Paganis, to whom the fortification treatise is dedicated, was one of the founders of the Knights Templar). After a wholly military education, Pagan entered military service at 12. His war experience included the sieges of Caen (1620), of St. Jean d'Angély, Cléric and Montauban (1621)—where he lost his left eye. He served at the sieges of La Rochelle (1627) and Nancy (1633) under Louis XIII, and followed the campaigns of Flanders and Picardy under Deville, who was the acknowledged first engineer of his time. In 1642 Pagan was blinded totally; while he could no longer attend sieges, he continued his mathematical studies with public success. His house became a kind of academy for scientists and writers—the military salon of Paris in the 1630s and 1640s—and his high social standing was confirmed when the King's own first physician attended his deathbed.

His publications include *Théorèmes géométriques* (Paris 1651, 1654), *Théorie des planètes* (Paris 1657), and *L'Homme héroique ou le prince parfait sous le nom du Roi* (Paris 1663). Pagan was the first engineer to be able to defend the crossing of the moat by placing enough cannon in the thickness of the flanks of the bastion. This treatise on fortification is the

summary of his experiences in war seen through a theoretical lens.

He postulates in his preface that if fortification were purely geometrical—implicitly criticizing some of the more theoretical publications—its rules could be perfectly demonstrable. But since fortification is dependent upon variables such as materials and the experience of its builders, its most essential maxims depend upon conjecture. All of Europe wonders at the relative weakness of today's fortifications, he writes; the strongest barely last six weeks, the best cannot be kept when there is an army nearby. His intention is to revolutionize these unsatisfactory practices.

The Newberry also owns a copy of the third edition of this treatise published in Paris in 1669 (worn brown calf, goldstamped spine, ex-libris Algernon Capell, Earl of Essex on title-page verso; 7.7 × 13.3 cm).

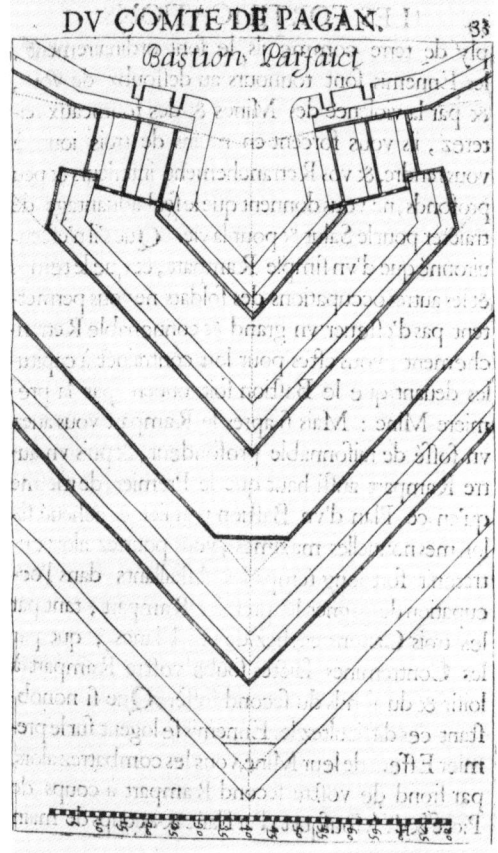

46 **Papillon, David,** 1581–c. 1655.
A Practical Abstract of the Arts of Fortification and Assailing.
London, R. Austin, 1645.

15.3 × 19.7 cm. Brown calf rebacked covers, blindstamped decoration and title on spine, ex-libris Henry B. H. Beaufroy, F.R.S. inside upper cover, ex-libris Fairfax of Cameron on verso of free endpaper.
x + 123 pages + i. Woodcut title-page with portrait of the author at 65 by Thomas Cross, preface, table of contents, 25 numbered woodcut illustrations (randomly placed in text, plates 24 and 25 are the first two, then 1–5 followed by 9, etc.), 105 × 100 to 189 × 150 mm.

First Edition. Dedication to Thomas Fairfax, General of the forces of the Honourable Houses of Parliament, is dated 1 January 1645.

REF: DNB
COPIES: DFo, NNE, NNUT-Mc.

An architect and military engineer, son of a captain and valet-de-chambre of Henri IV of France, Papillon came from a Huguenot family. His distinguished relatives included Almangue Papillon, Clément Marot's friend and the author of *Le Nouvel Amour,* and Antoine Papillon, a friend of Erasmus. He was brought up in London by relatives, after he escaped from France in 1588 in the wake of the St. Bartholomew massacre. His public offices included that of County Treasurer of Leicestershire (1642–1646), and Deacon of the French church in London. He fortified Gloucester for Parliament in 1646, and authored *The Vanity of the Lives and Passions of Men* (London 1651).

Papillon considers "defence and assailing" the most essential parts of the art of war. The motivations of the author are to procure common safety, to rectify the deformities of English fortifications (which "serve only as an object of derision to Forrainers that see them"), and to educate those of "the meanest capacities" interested in the practice of fortification. He criticizes the theoretical writings of Ward, Cruso, the author of the *Enchyridion,* and Norwood for obscuring the issues further. Among his topics are the true character of the engineer, the history of the art of fortification, the siting of the fortification, and the fortification of various regular polygons. Papillon recognizes the superiority of Dutch, French and Italian engineers, but does not write of Spanish and German fortifications because they were designed mostly by Italian engineers.

47 Perret, Jacques.
Des Fortifications et artifices, architecture et perspective.
n.p., n.d.

29.4 × 42.8 cm. Contemporary limp vellum, stained and scratched.
11 leaves. Engraved title-page and 23 double-page illustrations of fortified city plans and buildings, by Thomas de Leu, 250 × 256 to 534 × 412 mm.

First Edition. No dedication, but engraved view of siege of Paris, taken by Henri IV on 22 March 1594, on title-page.

REF: Brunet, Cockle, Manno, Architekt und Ingenieur, Miscellanea 12
COPIES: BN, BM (French and German editions of 1602, Frankfurt, Oppenheim 1613, Paris 1620), HAB (Paris 1601, French and German editions of Frankfurt 1602, Oppenheim 1613), MiU, DAIA, NjP, NN (Frankfurt 1602).

THE FIRST EDITION OF THIS "USELESS work, but still sought after for its splendid engravings by de Leu" (Brunet) was published in Paris in 1601. Perret was a Savoyard, possibly a Protestant from Chambéry, where he taught mathematics.

His book is divided into two parts: the first one illustrates ideal fortified cities, the second is focused upon religious and secular buildings inside the fortification walls. He dives right into the subject matter without preliminaries, the first page of his text describing the first fortress illustration, naming its parts and giving their dimensions. He is keenly interested in the geometry of fortification, allowing the drawings to speak for themselves. Perret's implicit message is that the pure geometry of the military defenses must find an equivalent counterpart in the orthogonal and radial street layout of the city. Thus, although Perret's text is not extensive, the use of emblematic elements in both plans and elevations and the architectural and structural contents of the illustrations are significant for architectural and military history. The last engraving illustrates in plan and perspective the massing, composition and interior space of a "temple." Comparative examination of fortification plans of regular four to 10-sided polygons shows alternative designs for walls, bastions and outworks. While the perspec-

tives are signed by both Perret and de Leu, the plans, although scaled, are not. Perret's designs seem to have been influenced by Pietro Cataneo's work and possibly by Philibert de l'Orme's architectural treatise.

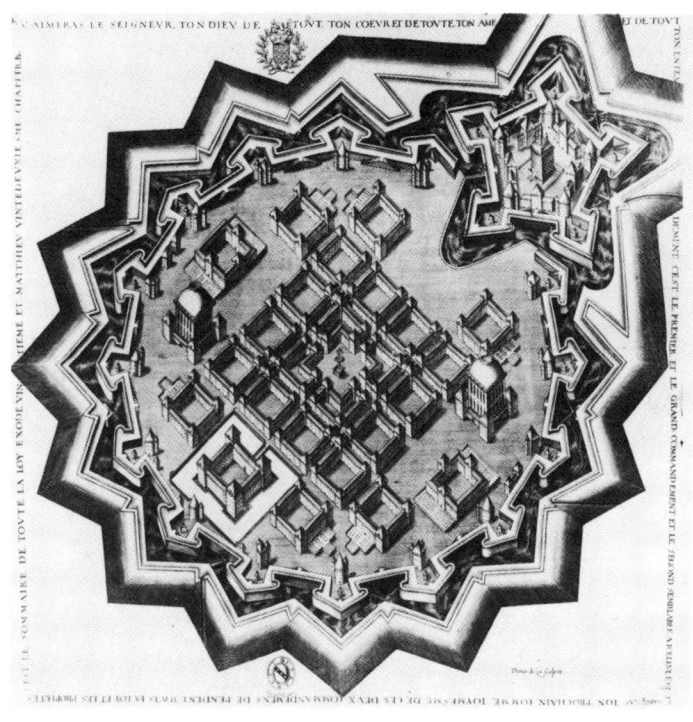

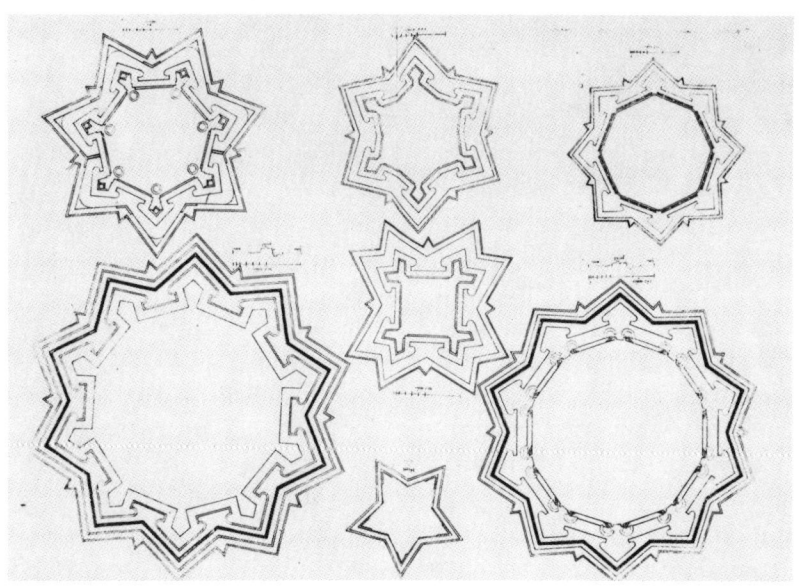

48 **Priorato, Galeazzo Gualdo, Count of Comazo,** 1606–1678.
Teatro del Belgio o sia descritione delle provincie del medesimo.
Frankfurt, 1673.

23.5 × 34.3 cm. Contemporary limp vellum, stained and yellowed, title in ink on spine.
xvi + 148 pages + vi. Engraved title-page: engraved coats-of-arms of 17 provinces, second title-page with engraved siege scene by Tob. Sadler, preface, list of illustrations, engraved map of the 17 provinces, engraved portraits of Louis XIV at p. 94, of Charles II at p. 99, of Louis de Bourbon, Prince of Condé at p. 106, of Henry, Count of Turenne and marshal of France at p. 114, Christopher, Bishop of Münster at p. 121, James, Duke of York, Jean, Count d'Estrée and Michel de Gruyter, Lieutenant Admiral of Holland at p. 122, engraved view of the crossing of the Rhine by the French army in 1672 and Paul Wirtz marshall of the United Provinces at p. 124, engraved naval battle at p. 127, Maurice Prince of Nassau at p. 128, Philip, Duke of Orleans at p. 130, William Henry, Prince of Orange at p. 132, Jan de Witt, Governor of Groote at p. 139, subject index, 138 plans of cities on 120 double-page engravings, 221 × 344 to 452 × 344 mm.

First Edition. Dedication to Prince Rinaldo d'Este, papal publication license dated 13 July 1673, Venetian publication license dated 19 June 1673.

REF: Michaud
COPIES: NjP (1683), MiU, NN, ScU, RPJcB (1683), OU (1683), IaU (1683).

As A HISTORIAN GUALDO PRIORATO enjoyed in his own lifetime a reputation not confirmed by posterity. He entered military service while very young, in Flanders, under the Prince of Orange. After being besieged in Breda, he became an ensign in the French army, and then went to England after 1626. He attended the siege of La Rochelle and then returned to Holland. He also visited Fez, and other African towns, served Wallenstein in the German wars, and then the Elector of Bavaria at the head of a Venetian regiment which was destroyed at the battle of Nördlingen. Although he eventually renounced military service, Priorato was knighted by the King of France and the Venetian Republic, the Pope made him a Roman noble, and Christina of Sweden named him first-gentleman, while the Emperor Leopold I appointed him counselor and historiographer.

Gualdo Priorato retired eventually to Vicenza, his home town, where he prepared his works for publication. Recognized as Count in his Venetian publication "copyright," he was the author of *Historia universale* (1652)—dedicated to Anne of Austria—in which he described the European wars of the previous ten years, *Istoria delle guerre degli imperatori Ferdinando II e III successe dal 1630 al 1640* (Bologna 1641, 3 vols.), *Istoria della vita d'Alberto Walstein* (Lyons 1643, Latin ed., Rostock 1668), *Vita e condizioni del Cardinale Mazarini* (Cologne 1662, translated into French, German and English), *Istoria di Leopoldo Cesare che contiene le cose più memorabile successe in Europa dal 1656 sino all'1670* (Vicenza 1670–74, 3 vols.), *Istoria di Ferdinando III imperatore* (Vienna 1672), *Istoria delle rivoluzioni di Francia sotto il regno di Luigi XIV dall'anno 1648 all'1654* (Cologne 1670, 2 vols., Pamplona 1720), *Vies des Princes de Savoie, Theatre des hommes illustres d'Italie,* and *Le Guerrier prudent et politique.*

The *Teatro*—part of *Description des*

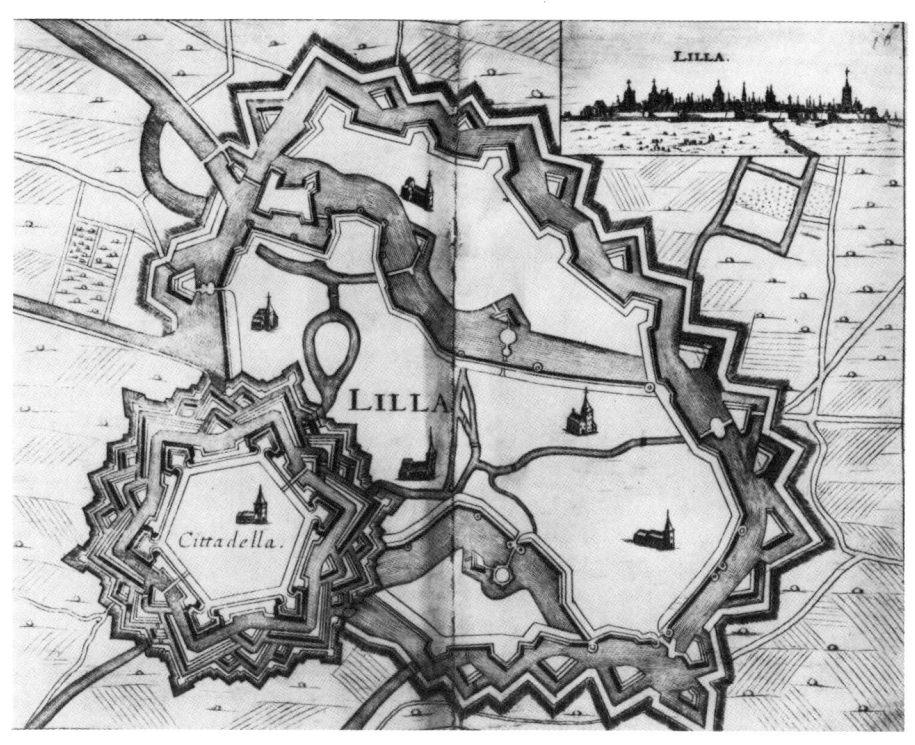

principales villes d'Allemagne, des Pays-Bas et de l'Italie—is not actually a fortification treatise, but rather a description of Belgica (the Roman name for today's Belgium and Netherlands), which in the seventeenth century was divided into Catholic and Protestant parts, at war with one another. He describes each of the Catholic and Protestant principalities, including their principal cities, forms of government, economic systems, and diplomatic relations, concluding with a description of the war waged by Louis XIV against the United Provinces in 1672. While the text is more descriptive than the pedagogical and analytic texts of most fortification treatises, this work provides some of the same materials by illustrating the fortifications of a great number of cities. It could well have served the officers of the attacking army in the planning of their siege strategy. It is important to note that the internal layout of the individual cities is not shown in their plans, where the emphasis is upon the fortification enclosure and the outworks, as well as on the surrounding site and bodies of water.

49 Rossetti, Donato, 1633–86.
Fortificazione a rovescio.
Turin, Bartolomeo Zappata, 1678.

23 × 36 cm. Worn brown calf, "di Antonio Orsetti" on free endpaper, speckled edges.
xii + 211 pages + i. Table of contents, 58 engraved figures of fortification studies, (Figs. I–II, V–X repeated twice, XI–XIV repeated three times), 93 × 162 to 425 × 357 mm, Fig. 52 engraved by Joannes Fayneau.

First Edition. Dedication to Maria Giovanna Battista, Duchess of Savoy and Regent, and to Vittorio Amedeo II, Duke of Savoy, is dated 15 January 1678, license to publish granted by local Inquisitor on 17 August 1677.

REF: Marini
COPIES: N.

DONATO ROSSETTI WAS A THEOLOGIAN from Leghorn, reader in philosophy at University of Pisa and professor of mathematics for the duke of Savoy.

The treatise is divided into two dialogues in which the author sets forth his method of fortification "in reverse." This term is used in order to characterize a fortification in which the moat is narrowest at the tip of the bastion projecting further into the countryside in front of the curtain wall. Since the moat follows the placement of the outworks in front of the curtain wall this may well be only an academic term for an already existing fortification method. Fig. LII clarifies the intentions of the author: in effect he is incorporating the outwork, which becomes a modified bastion, in front of the curtain wall but inside the wet moat thus obliging the besieger to modify the traditional strategy of attacking the bastion. The work of Blaise Pagan is mentioned often, and there are also numerous references to Manesson Mallet's treatise. Marini praises Rossetti, attributing to him the revival of Antonio de Sangallo's Roman bastion.

FIG. LII. pag. 192.

Joannes Fayneau Sculpsit Taurini

50 Ruggiero, Pietro.

La militare architettura overo moderna fortificatione.
Milan, Lodovico Monza, 1661.

19.1 × 25.2 cm. Half limp vellum, patterned paper boards.
xii + 238 pages + vi. Woodcut title-page, preface, 54 numbered woodcut illustrations, but #39 missing, 125 × 56 to 350 × 248 mm, subject index.

First Edition. Dedication to John of Austira; license to publish released by the Inquisition on 3 February 1661.

REF: Guarnieri
COPIES: DLC, MiU, NcD, NN.

Ruggiero seems to have come from Burgundy, and held the rank of Captain and engineer in the army of the King of Spain whom he served in Flanders and in Germany.

The treatise is divided into four parts. Book one is about the principles of mathematics and geometry, both of which are fundamental for the practice of military architecture. Book two—the longest—deals with fortification itself, considering themes such as site of the fortress, the form of the fortress, variations in bastion design (as seen in the theoretical work of Pagan, and in built examples), and the role of the citadel. Book three discusses the various rules adopted for the garrisoning of a fortress, such as the number of soldiers, gate guards, siege strategies, assault, mining, and countermining. Book four focuses on "military art", the three main activities of the army: marching, lodging and fighting and the topics range from provisioning to ammunition to the layout of a camp.

51 **Ruse, Heinrik (Baron von Rysenstein or Rusensteen)**, 1624–1674, and Gerhard Melder.
Praxis fortificatoriae oder Kunst-gründige Anweisung, wie und welcher Gestalt die heut zu Tag gebräuchliche Fortificationes gar mercklich verbessert und verstärcket. Osnabrück, Johann George Schwänder, 1664.

17.5 × 28.5 cm. Contemporary limp vellum, inked title on spine, half title-page.
x + 135 pages + iii. Preface, subject index, 83 woodcut diagrams, 78 × 76 to 348 × 286 mm.
First German Edition. Dedication to Ernst Augustus, Prince-Bishop of Osnabrück.

REF: Architekt und Ingenieur, Guarnieri, BWN
COPIES: HAB (1654, Frankfurt 1666, 1670).

In military service in the United Provinces from the age of 15 for four years, Ruse subsequently served the French monarchy under Turenne, the Republic of Venice under Lunardo Foscolo, and several German princes. His contribution is outstanding because he renewed—together with Minno van Coehoorn—the old Dutch method of fortification. Among the positions he held were city-engineer of Amsterdam and designer of the fortifications of Berlin, Spandau and Kalkar; from 1661 he was General Inspector of Fortification for Frederick II (1609–1670) King of Denmark, for whom he built one of the most beautiful extant seventeenth-century fortresses, the citadel of Frederikshavn in Copenhagen, finished in 1667. He was named a baron for his Danish service.

This is the first German edition of his treatise (Frankfurt 1666 is the second), first published in Dutch as *Versterckte Vesting* (1654), and then in English as *The Strengthening of Strongholds* (1668). These publications ensured his renown as both theoretician and practicing military architect, even though, according to Guarnieri, his co-author Melder plagiarized and altered Freitag's *Architecture militaire* in the composition of this joint publication, while Ruse paraphrased Pagan in his contribution to this treatise. Gerhard Melder was a military builder from Utrecht, where he also taught fortification. His best-known work, titled *Korte en klare Instructie van regulare en irregulare fortificatie, met hare buytenwercken, te gebruycken defensive en offensive*, was published in Utrecht in 1658, with a German edition in Osnabrück in 1661. Melder and Ruse held opposing views on fortification, and debated acrimoniously in print; their joint treatise is a clever publisher's trick.

The book is divided into four sections: the strengthening of modern fortifications, instructions for new fortification, offensive and defensive war, and the mathematical principles of fortification.

52 Sardi, Pietro, 1560–after 1642.
Corona imperiale dell'architettura militare divisa in due trattati. Il primo contiene la teorica. Il secondo contiene la pratica.
Venice, Barezzo Barezzi [at author's expense], 1618.

25.1 × 36.2 cm. Contemporary limp vellum, two pairs of ties removed, large worm-hole in lower cover, ex-libris Baron Horace of Landau inside upper cover, stamp of Bibl. Galeotti on title-page.
xxviii + 299 pages + i + 83 pages + i. Title-page with portrait of the author at age 58 engraved by Gaspar Grispoldi, introduction, subject indices for part one and part two, two engraved part title-pages, 37 engraved illustration of bastions and fortifications in part two (two of #2, 3 of #3, 5 of #5), 198 × 145 to 200 × 228 mm; 4 engraved figures of geometrical diagrams in appendix on principles of geometry, 198 × 146 mm.

First Edition. Dedication to Bartolomeo Lomellino, Giovan Domenico Pallavicino, Giacomo Cattaneo and Giorgio Doria dated 24 November 1618; Venetian license to publish dated 10 November 1618.

REF: Cockle, Architekt und Ingenieur
COPIES: BM (1618, French Frankfurt edition of 1623), Bodleian (French Frankfurt edition of 1622), HAB (German, Frankfurt 1622—two copies, Italian edition 1639), BN (1623), MB, CU, NNC, NjP, MiU (German, Frankfurt 1623), NN (German, Frankfurt 1623).

Pietro Sardi was an author of Roman origin, responsible for the *Corno dogale della architettura militare* (1639) and the *Discorso sopra la necessità et utilità dell'architettura militare* (1642). Like his *Discorso,* the text of the *Corona imperiale* is generously interlaced with quotes in Latin from illustrious ancient Roman writers such as Livy, Vegetius, Suetonius and Polybius. His books appealed to a learned elite (he says at the beginning of his *Discorso* that military architecture should be reserved for the elite) whom he flattered into being interested in strategic fortification by evoking the example of famous military leaders from antiquity.

The *Corona* is divided into two parts, a theoretical section and a practical one. The first part is further subdivided into seven chapters: purpose of fortification, sites, assault methods, form of the fortification, construction materials, provisioning of the fortress and the defense of the fortress. In the second part Sardi demonstrates through diagrams the principles discussed in the first. While the first part is intended for those who know Latin and could enjoy the range of Greek and Latin authorities quoted by Sardi, the second part is for simpler souls more directly interested in the applied science of fortification.

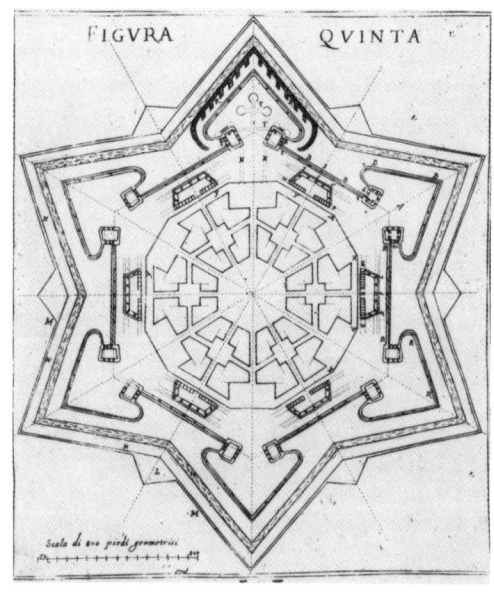

53 Sardi, Pietro, 1560–after 1642.
Corno dogale della architettura militare.
Venice, Giunti, 1639.

21.5 × 31 cm. Quarter calf, title goldstamped on beige panel, beige and black speckled paper boards, half-title, "Michaelij Arcangeli Seviati 1670" in ink on title-page.

xxviii + 220 pages. Engraved title-page with portrait of the author at age 79, preface, subject index, introduction, 44 engraved plates (24 numbered figures [#4 and 5 missing] in Book Two; 15 engraved diagrams of Dutch fortifications in Book Three, 5 diagrams of Roman military camps), 118 × 132 to 386 × 282 mm.

First Edition. Dedication to Francesco Rizzo, Doge of Venice, is dated 15 February 1638.

REF: Cockle
COPIES: BM, BN, MB, MH, NN, MiU, DFo.

PUBLISHED TWENTY YEARS AFTER THE *Corona Imperiale*, *Corno dogale* is divided into seven chapters: definition of military architecture, siting, provisioning and defense of the fortress, Dutch fortification methods, Roman military camps, the lodging of the army and the fortification of its camp, battle strategy, and algebraic exercises. As in the earlier work, Sardi quotes Latin sources extensively especially in the first chapter. The long introduction, where Venice is presented as analogous to Imperial Rome, explains the title of the book, which praises the Republic as the only part of Italy free from foreign rule.

54 Scamozzi, Vincenzo, 1552/7–1616.
L'idea dell'architettura universale divisa in X libri.
Venice, by the author, 1615.

23.2 × 34 cm. Eighteenth-century brown rebacked calf, blindstamped covers with traces of gold, gold-stamped title on spine, "Jacobi Joye" on title-page.
xvi + 352 pages + xxxiv + xii + 370 + xxii. Engraved title-page, 5 part title-pages, preface, 2 tables of contents, introductions to Part one and to Part two, 40 mixed engraved and woodcut plates in Part one, 188 × 279 to 400 × 279 mm, two subject indices, 46 mixed engraved and woodcut plates in Part two, 188 × 280 to 400 × 280 mm.

First Edition. Part one dedication to Maximilian, Archduke of Austria, is dated 6 August 1615; Part one, Book Two dedication to Carlo Emanuele I, Duke of Savoy, is dated 15 August 1615; Part one Book Three dedication to Maximilian, Duke of Bavaria, is dated 20 August 1615; Part two dedication to Cosimo, Grandduke of Tuscany, is dated 6 August 1615; Part two Book Seven dedicated to Francesco Maria II, Duke of Urbino, is dated 8 September 1615; Part two Book Eight dedicated to the Academici di Vicenza, is dated 8 December 1615.

REF: Wiebenson, Encyclopedia of Architects, de la Croix
COPIES: MH, GAT, CtU, DDO, OKentU, NjP, NBuU, DeU, MnU, ICU, CtY, MoU, NcRS, NN, TU, NcU, OO, IaU, TxU, NcD, DLC, WaU, OrCS, OrU, CaBVaU, MtBC, KU, NB, NIC, MoSU, PSt, Cu.

A NATIVE OF VICENZA, SCAMOZZI studied first with his father, an architect, and frequented the Accademia Olimpica. He was the executor of Palladio's unfinished Teatro Olimpico in Vicenza, having worked with the great architect as his assistant in Venice. He travelled to Rome and Naples in 1579, and was in Rome among the Venetian delegation sent to congratulate Sixtus V. His journey in 1600 through France, Lorraine, Germany and Hungary in the suite of Venetian ambassadors is documented in a MS diary (ed. Franco Barbieri, *Taccuino di viaggio da Parigi a Venezia*, Rome, 1959). Scamozzi is among the sixteenth-century architectural theorists who also produced a large body of architectural works. These include the Procuratie Nuove (1581–1598) in Venice and the design of the small town of Sabbioneta, on which his claim to interest as an urbanist must be based.

He began to work on his *Idea* in 1591 having started its planning in 1583 when he published his *Discorsi sopra l'antichità*

di Roma (Venice 1583). His original plan was to write twelve books, which were then reduced to ten; of these only six were first published. They cover the history and theory of architecture, fortification and town planning, palaces and villas, the Orders, building materials, and construction. Recent historians consider its influence to have been negligible despite the testimony of numerous subsequent reprints (1691, 1714, 1835, 1838; Dutch, Amsterdam 1640; German, Nuremberg 1678).

Book two of part one, dedicated to Carlo Emanuele I, is a discussion of the urban site which includes the dimensions of great cities, the quality of ports, the layout of streets and principal buildings within the city, and the form and constituent elements of fortification as well as their design and construction.

Scamozzi is the last treatise writer to integrate both civic and military architecture in one study according to de la Croix. He does provide an interesting literary and graphic prescription for the "universal" form of the city (Part One, 164–70). His instructions cover the entire range of urban considerations from the width of streets (which should not be wider than the height of flanking buildings in order to maintain optimal temperatures), and their pavement (which should not be of stone because it is too noisy), the design of the five proposed squares of the city, the fortified walls, to the perimeter of the city (3 miles long and surrounded by a military street).

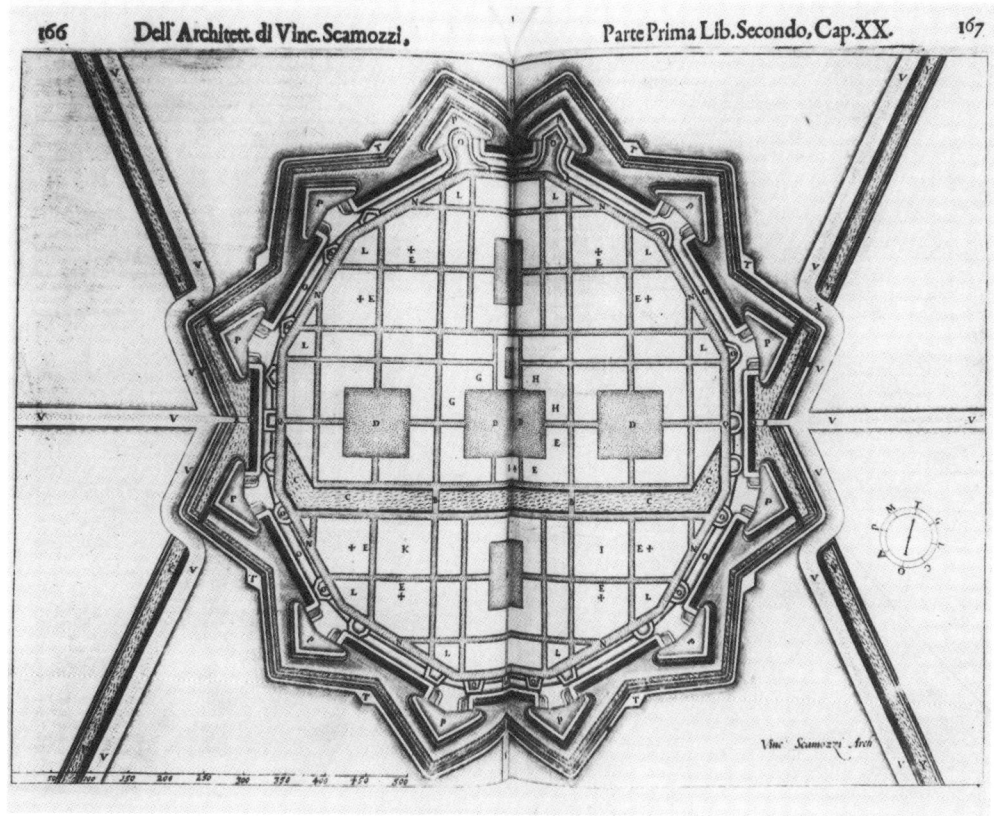

55 **Speckle, Daniel,** 1536–1589.
Architectura von Vestungen. Wie die zu unsern Zeiten mögen erbawen werden.
Strasbourg, Bernhart Jobin, 1589.

25 × 36 cm. Blindstamped, tawed pigskin over wooden boards, one of two brass clasps missing, frequent marginalia in Italian, in ink and in lead.
viii + 112 leaves. Engraved title-page, introduction, table of contents, 24 woodcut diagrams in text, 148 × 50 to 158 × 158 mm; 20 double-page engraved plates of fortification studies and actual fortresses in plan and perspective, 402 × 300 mm, 1 single-page plate, 192 × 308 mm.

First Edition. Dedication to Julius, Prince of Brunswick and Lunenburg, is dated 1 February 1589.

REF: Architekt und Ingenieur, Cockle, ADB
COPIES: BM (1589, 1599), BN (1589, 1599), HAB (colored woodcuts and engravings), Strasbourg (MS 1583), N, NN, MoU, MiU (1608), DFo (1608), NNC (1608).

SPECKLE WAS AN ARCHITECT FROM Strasbourg; he was named city-architect of his home town in 1576. Until then he had worked as a military engineer in Düsseldorf and Regensburg (1567), in Vienna (1569) called there by Carlo Tetti, and in Bavarian military service (1574). Speckle's travels through Germany, Denmark, Sweden, the Netherlands and Hungary are documented in his numerous autograph drawings of city plans and fortifications; he was actively engaged in the fortification of Vienna.

His treatise is a standard reference work for sixteenth-century German fortification methods, fully developing the suggestions made by Dürer earlier in the century and marrying them to the most recent innovations in military architecture. It was the first treatise in German to deal with the polygonal bastioned fortification trace and to compare citadels, ideal cities, mountain fortresses and city fortifications to one another. Speckle rebelled against the traditions of Italian late-Renaissance military architecture (which disdained other ideas and practical men unable to read the Latin authors), and the monopoly on fortifications enjoyed by Italian authors.

This treatise is divided into three parts: construction of principal elements of the fortification, problems of site (mountain and plain towns), and maritime and river towns. Some of his problematic solutions resemble those by Errard; as the number of the sides of the polygon increase, the length of the curtain wall decreases, placing bastions closer to one another. His scenographic perspective drawings are breathtaking, as are the views of mountain fortifications identified in the 1599 edition as fortresses in the vicinity of Strasbourg, but equally inspired by Dürer's illustrations of castles.

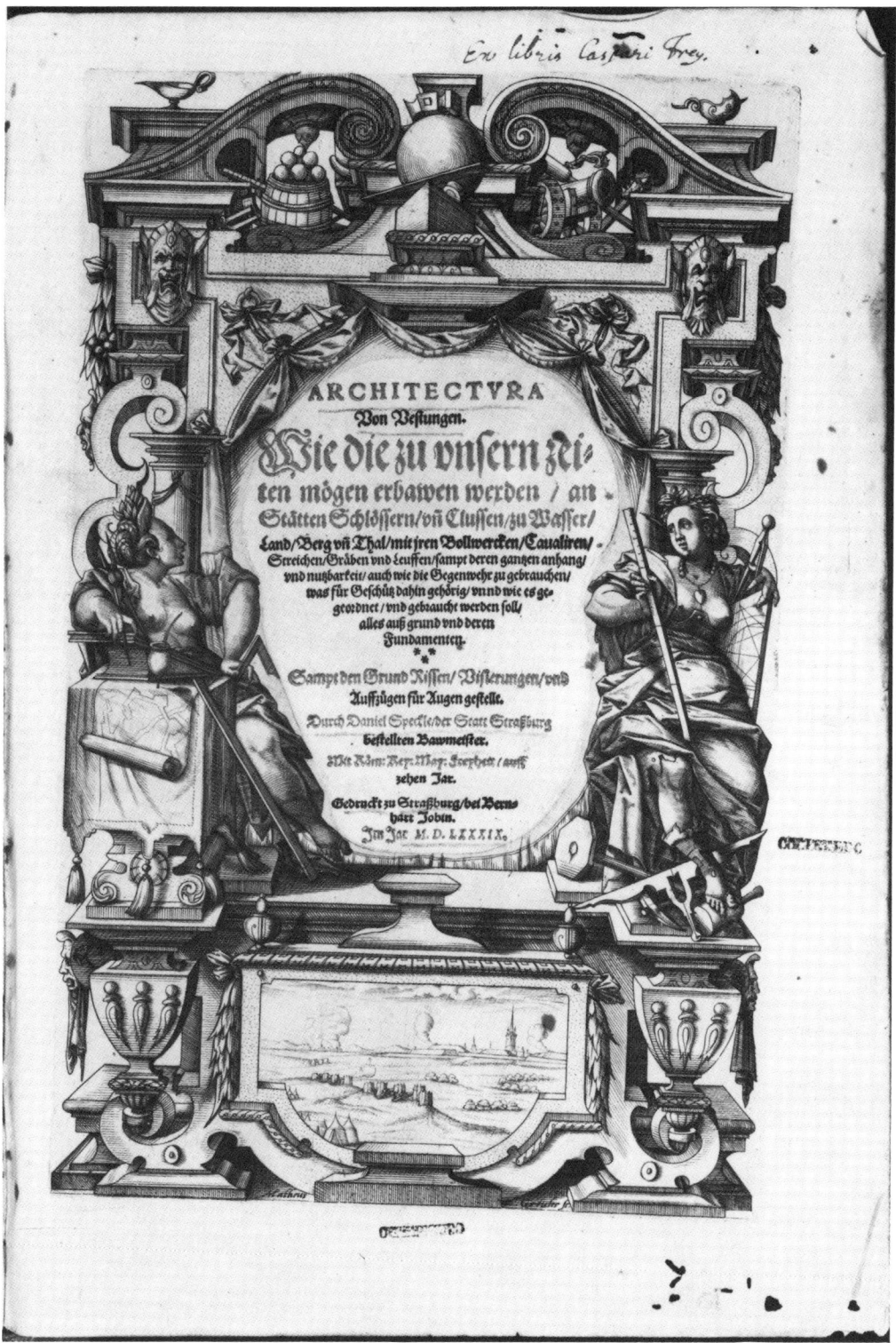

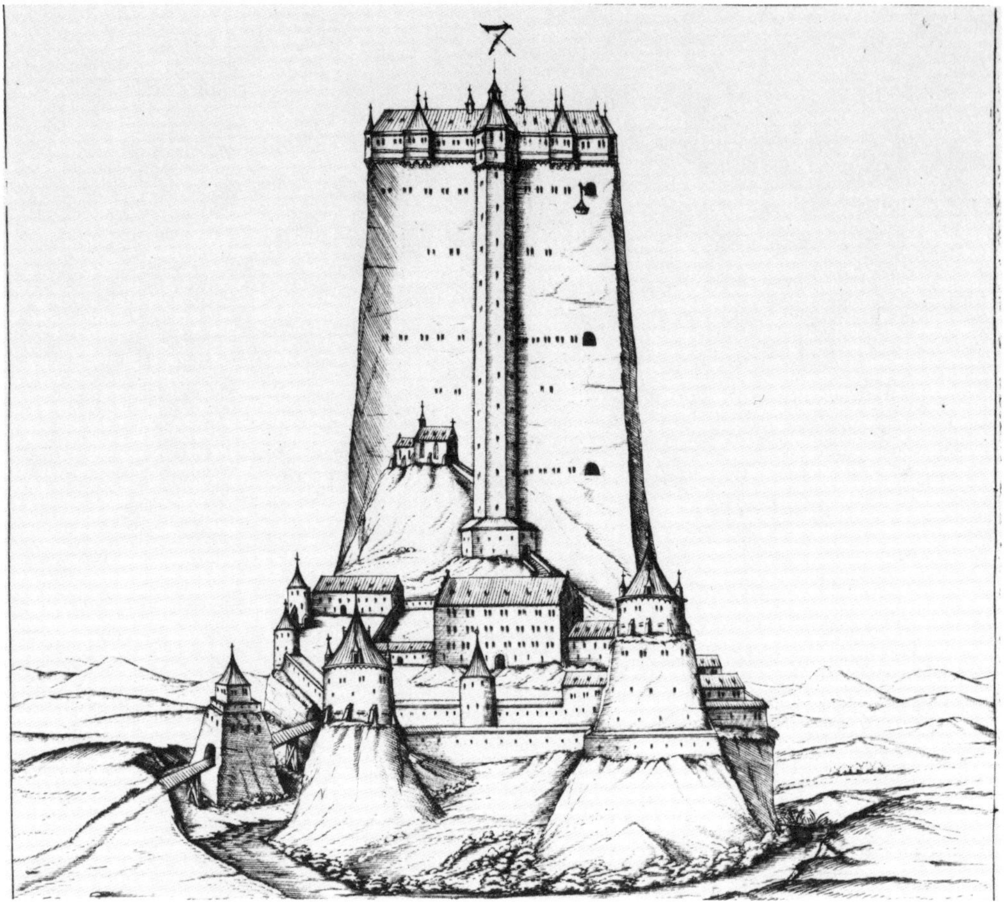

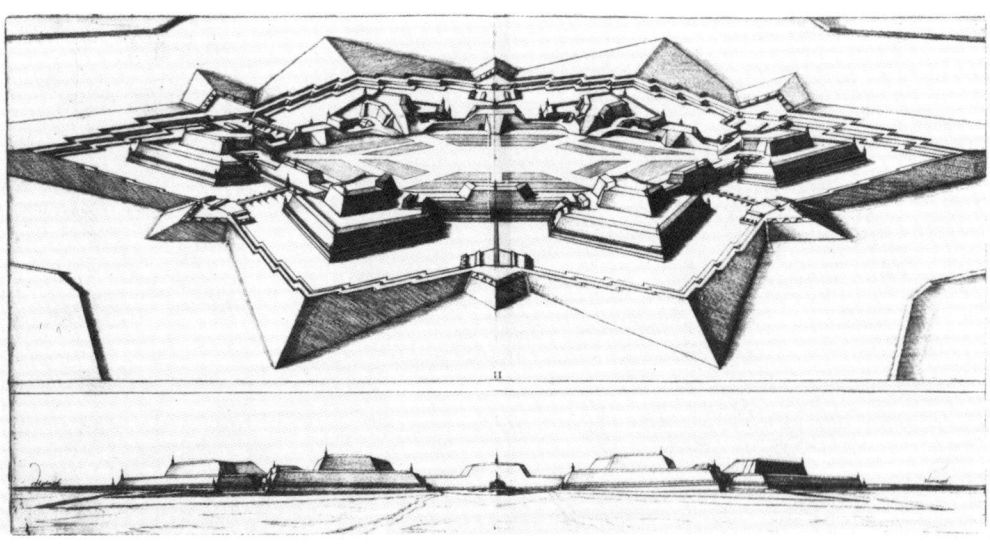

56   S[teed], J[ohn].
Fortification and Military Discipline in Two Parts.
London, Robert Morden, 1688.

10 × 16.7 cm. Half calf, brown paper boards, goldstamped title on red morocco panel.
viii + 132 pages + iv. Engraved and water-colored frontispiece, preface by the publisher, table of contents, 33 (rather than the 54 copper plates announced on the title-page) engraved full-page illustrations of fortress plans and fortification studies (plate III missing), 98 × 167 to 220 × 182 mm.
First Edition.
COPIES: CSmH, NN, SS.

THE FIRST PART OF THIS TREATISE DEALS with the principles and practice of all manner of fortifications, regular and irregular, as used by the Dutch, English, Italian, German and French engineers; the dimensions and measures of the ramparts, parapets, and moats; the delineations of ravelins, halfmoons, hornworks, crownworks, and citadels; and also the making of batteries, approaches, trenches, mines and assaults. The second part treats the rules for the exercise of horse and foot soldiers, and instructions and observations related to the whole art of war.

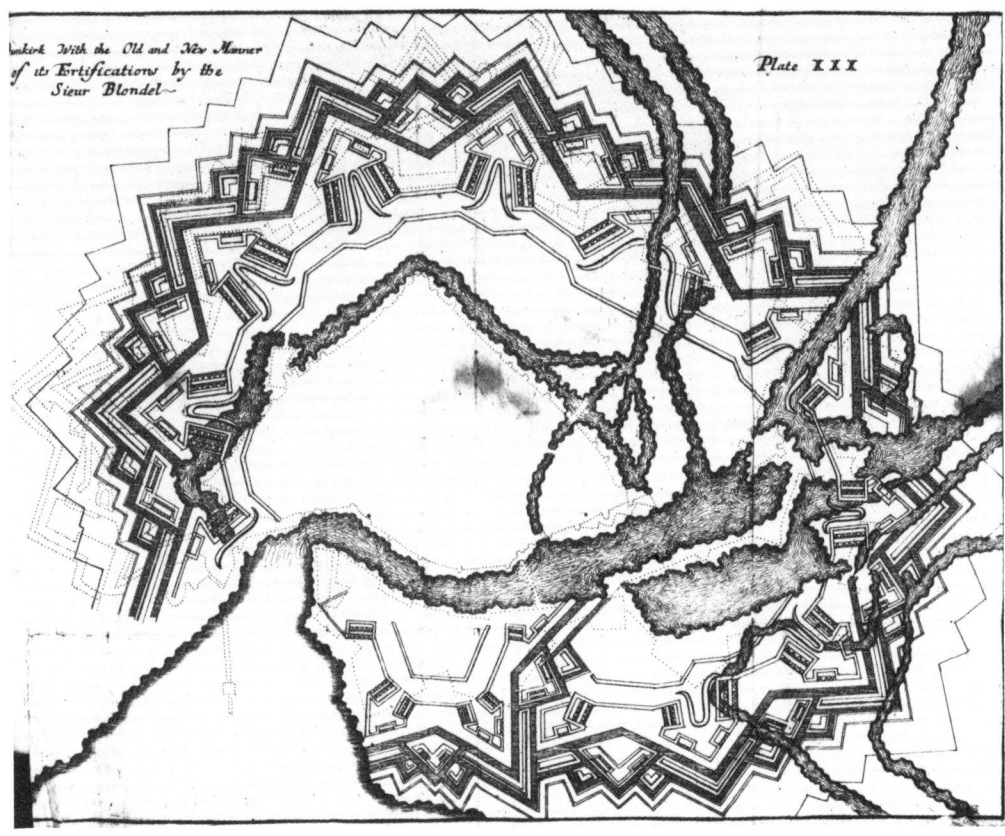

57 **Stevin, Symon,** 1548–1620.
Nieuwe maniere van sterctebou door spilsluysen.
Rotterdam, Ian van Waesberghe, 1617.

iv + 59 pages + iii. Table of contents, 38 woodcut diagrams and plans of fortifications, 137 × 70 to 300 × 408 mm.

First Edition. Dedication to the Estates General of the United Provinces is dated 21 December 1617.

REF: Cockle
COPIES: BM, BN (French, Leiden 1618), IaU, NIC, MH (French, Leiden 1618), NjP (French, Leiden 1618), MiU (1617; French, Leiden 1618), MnU (French, Leiden 1618), NNC (French, Leiden 1618).

A NATIVE OF BRUGES, STEVIN IS KNOWN as the "Father" of modern statics. "*Stevinus,* that great mathematician and engineer," is immortalized in Laurence Sterne's *Tristam Shandy* as the inventor of a wind-driven chariot and the author of Uncle Toby's favorite work on "the science of fortification" (II, xii, xiv). He was not only a mathematician and theoretician of military architecture, but from 1617 was also engineer and surveyor in the service of Maurice, prince of Orange. After 1571 he visited Prussia, Poland, Sweden and Norway, settling in Holland. He matriculated at the University of Leiden in letters in 1583. His publications include a book on calculating banking interest and double-entry book-keeping. His MS papers are preserved at the Royal Library in The Hague. Stevin took part in the siege of Jülich in 1610, and used his experience in the formulation of his books. He innovatively bound together castrametation with siege strategy, tactics and fortification design, creating a scientific axis between them. Furthermore, the use of sluices made a virtue out of the excessive water channels of Holland, by placing them at the center of a new national fortification method.

58 **Stevin, Symon,** 1548–1620.
Castrametatio dat is Legermetig.
Rotterdam, Ian van Waesberghe, 1617.

19.4 × 28.8 cm. Blue marbled paper boards, yellow edges.
vi + 55 pages + i. Title page with engraved decoration, engraved coat-of arms of Orange family and motto of the Garter (*Honi soit qui mal y pense*), 14 woodcut diagrams in text, 23 × 98 to 63 × 144 mm.
First Edition. Dedication to Maurice, Prince of Orange-Nassau, and the Estates General of the United Provinces is dated 4 November 1617.
REF: Architekt und Ingenieur, BN, Michaud
COPIES: HAB (1617, 1633, French 1618), BN (1618), MH (French 1618), NjP (French 1618), MiU (French 1618), CtY, NIC.

*C*ASTRAMETATIO IS DIVIDED INTO TWO parts: the 3 chapters of Part One deal with the camps of the Estates General, while the one chapter of Part Two provides an analysis of ancient Roman encampments. Stevin describes and defines castrametation, provides a list of what ought to be accommodated in a camp, and discusses how to survey a camp and its long-lasting form. The Newberry copy is bound with his *Nieuwe maniere van sterctebou door spilsluysen.*

59 **Stevin, Symon,** 1548–1620.
Festung Bawung das ist kurze und eygentliche Beschreibung wie man Festungen bawen.
Frankfurt, Wolfgang Richter and the widow of Levin Huls, 1608.

15.4 × 18.9 cm. Bound in reused parchment from a legal manuscript of the 14th century, four leather ties, one removed.
viii + 132 pages. 32 woodcut diagrams, 20 × 25 to 201 × 189 mm.
First German Edition. Dedication to Valentin Sebitz is dated 1 November 1607.
COPIES: ICJ (1623), MiU CtY (1623), MnU (1623), NNC (1623), PPT (1623).

60 Stone, Nicholas, 1586–1647.
Enchiridion of Fortification or a Handfull of Knowledge in Martiall Affaires.
London, Richard Royston, 1645.

10 × 16.7 cm. Marbled paper boards, red morocco spine, goldstamped title on spine.
viii + 70 + xviii. Engraved frontispiece, poems to the author, introduction, 20 engraved full-page, folded-in illustrations, 168 × 166 to 175 × 166 mm, glossary of terms, subject index.

First Edition.

REF: DNB
COPIES: DLC (1669), MiU, CLU-C, CtY, NNC, MH, MnU.

A MASON, SCULPTOR AND ARCHITECT, Stone was the son of a quarryman apprenticed to Isaac James in London. He worked in Amsterdam before 1614 under Pieter de Keyser (the son of the more famous Hendrik de Keyser). Employed by James I, Stone carried out several designs by Inigo Jones, including the Banqueting House and the gates of the Physick Garden at Oxford. By 1619 he was master-mason to James I, while a patent letter of Charles I in 1626 named him architect of Windsor. He is best known for his monuments (John Donne's at St. Paul's is by Stone) and his connection with Rome through his second son, also Nicholas, who worked in Bernini's workshop. An account book of Stone's is preserved at the Soane Museum.

The author intends to demonstrate through "rule and figure" how to fortify regular and irregular bodies, how to run approaches, how to pierce through a counterscarp, how to spring a mine, and how to solve many other problems of fortification.

61 Sturm, Leonhard Christoph, 1669–1719
Architectura militaris hypothetico-eclectica, oder gründliche Anleitung zu der Kriegs-Baukunst.
Nuremberg, Peter Conrad Monath, 1720.

18 × 22 cm. Half vellum, marbled paper boards.
xviii + 144 + 23 + i. Engraved portrait of the author, introduction, table of contents, appendix, 42 full-page engraved diagrams of fortifications and bastions folded-in, 356 × 220 to 362 × 298 mm.

Third Edition. Dedication to Eugene, Prince of Savoy and Piedmont, by the publisher.

REF: Architekt und Ingenieur, Wiebenson, Michaud
COPIES: HAB (Nuremberg 1702, 1719, 1735), MnU (1755).

A CELEBRATED ARCHITECT, STURM WAS born in Altdorf where his father was professor of mathematics, philosophy and the "restorer" of physical sciences in Germany. Sturm studied in Leipzig and subsequently taught mathematics at

Frankfurt am Oder, where he was responsible for construction supervision for the Duke of Mecklenburg. Sturm influenced military architectural practice less through his buildings than through his publications. He taught mathematics and both civil and military architecture at the Wolfenbüttel Academy from 1694 to 1702, in 1711 occupied the post of construction supervisor in Schwerin, and in 1719 he was appointed to an equivalent post in Blankenburg, where he died.

His numerous publications include an edition of Nicolas Goldman's treatise (Augsburg 1721), and his own works *Der geoeffnete Ritter* (Hamburg 1700), *Introductio ad architecturam militarem* (Frankfurt 1703), *Warhafftiger Vauban* (Frankfurt 1703), *Le veritable Vauban se montrant au lieu du faux Vauban* (The Hague 1708), *Freundlicher Wettstreit der Französischen, Holländischen und Teutschen Krieges-Baukunst* (Augsburg 1718), and *Neue Manier zu befestigen* (Hamburg 1718). At the Newberry is preserved also his *Architectura militaris hypothetica et eclectica, das ist: eine getreue Anweisung wie man sich der gar verschiedenen . . . Befestigungs Manieren . . . bedienen könne* (Nuremberg, 1702) which is more modest than this 1720 edition but contains some of the same illustrations.

The German counterpart of Ozanam, Sturm's publications provided the foundation for a history of post-cannon fortification by systematically describing and comparing the innovations and errors of sixteenth and seventeenth-century military architects practicing in the four principal participating regions of western Europe (France, Holland, Italy, and Germany).

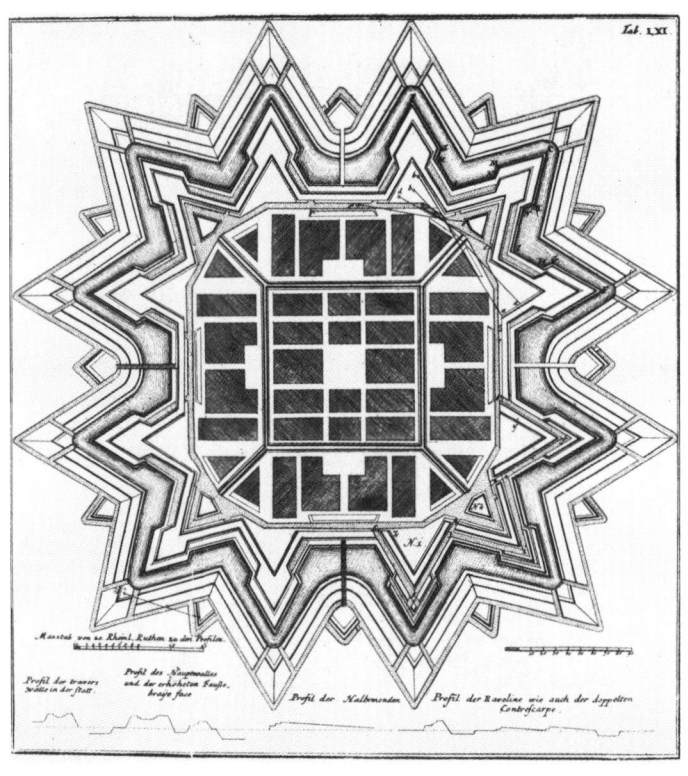

Leonhard Christoph Sturms
ARCHITECTURA MILITARIS HYPOTHETICO-ECLECTICA,

Oder

Gründliche Anleitung

zu der

Kriegs-Baukunst,

Aus denen Hypothesibus und Erfindungen derer meinsten und besten

INGENIEURS

dargestellet.

Nürnberg,

Verlegts Peter Conrad Monath. 1720.

62 Tacquett, Andrew, 1612–1660.

Military Architecture or the Art of Fortifying Towns; Together with the Ways of Defending and Besieging the Same.
London, 1672.

19.5 × 29.5 cm. Brown calf; this copy is bound with *Military and Maritime Discipline* and *The Compleat Gunner* (London, 1672).
iv + 68 pages. 16 engraved figures, 11 full-page, 157 × 80 to 341 × 294 mm.

First English Edition. Dedication to Aubrey de Vere, Earl of Oxford, by John Lacy, translator from the Latin.

REF: BNB
COPIES: MiU (Antwerp 1668), NNC, CLU-C.

A JESUIT MATHEMATICIAN, TACQUETT was born in Antwerp where his father was a wealthy merchant. He studied at the Jesuit college in Antwerp, entered the order in 1629, and spent the two years of novitiate at Malines, followed by four years of logic, physics and mathematics at Louvain. Following Jesuit custom, he taught subjects such as Greek and poetry for five years in the company's college at Bruges. After four more years of theological studies at Louvain he was given the chair in mathematics at the same university, but was recalled to Antwerp to teach the young Duke of Enghien, son of the Grand Condé. His correspondence with the scientist Christiaan Huygens, who compared him to Pascal and Liebniz, was arranged by the painter Daniel Seghers. Tacquett's published books include *Des Cylindres et des anneaux* (1650) and *Opera mathematica* (Antwerp 1668, dedicated to Cardinal Rospigliosi).

This brief military treatise is part of the *Opera mathematica* (its length ran to about 45 pages in the first posthumous edition); an anomaly among Tacquett's other works, it may have been composed for the instruction of the Duke of Enghien.

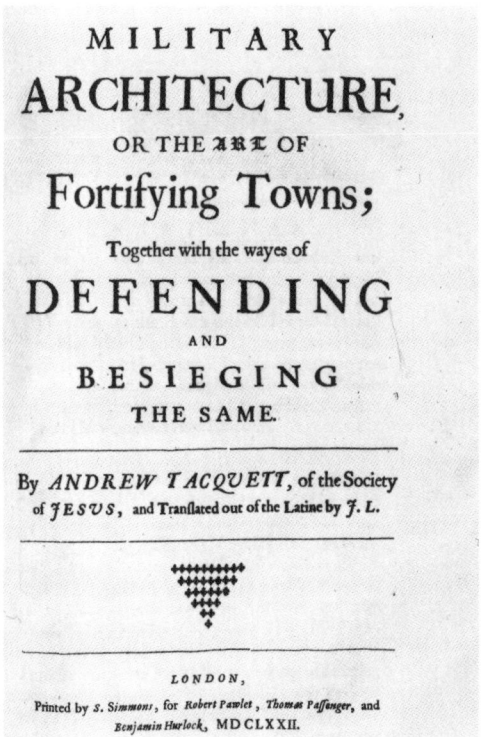

63 Tartaglia, Niccolò, c. 1499–1557.
La nova scientia.
Venice, 1537.

15.5 × 20.6 cm. Blue paper boards.
iv + 32 leaves (three restored). Woodcut title-page, table of contents, 32 woodcut figures, 77 × 45 to 152 × 206 mm.

First Edition. Dedication to Francesco Maria della Rovere, Duke of Urbino, is dated 20 December 1537.

REF: Cockle, Manno, Michaud
COPIES: BM (1537, 1550, 1558, 1606), HAB, MiU (1550, 1558), ICU (1558), NN (1537, 1558, 1583), NNC (1537, 1558), NjP (1537, 1558), DFo (1537, 1583), ICJ (1537, 1583), InU (1537, 1562, 1583), NIC (1537, 1558), OkU (1537, 1558), NSyU, DLC (1550), TxU (1550), MH (1550), CtY-M (1550), NcRS (1550), CtY (1558), WU (1558), MnU (1558), MNS (1583), MB (1583).

Although the table of contents lists five books, only three were actually published in this and the subsequent editions of the *Nova Scientia*. On the extraordinary title-page Tartaglia is shown explaining his discovery of the hyperbolic curve in the garden of geometry to the personifications of the sciences and the liberal arts (only Philosophy is missing, though watching the proceedings from a higher distance).

A native of Brescia, Tartaglia was orphaned at five and a beating he received in 1512 from the soldiers of Gaston de Foix left him with a speech impediment. Self-educated, he taught geometry and mathematics in Verona, Vicenza, Brescia and Venice, where he also translated Euclid. Tartaglia was cheated by his friend Cardano, who published Tartaglia's solution for the square root as his own. Their disputes, published and debated in public, diffused their scientific discoveries and promoted the development of mathematics. Tartaglia was one of the first to apply mathematics to artillery and military art.

Among his publications are *Euclide* (Venice 1543), the first Italian translation of Euclid; *General trattato de numeri e misure* (Venice 1556, 1560); *Trattato di aritmetica* (Venice 1556, Paris 1578). In this treatise he presents his discovery that the trajectory of a cannon-ball is curved, given that it is the result of the conflicting powers of impetus and gravity.

64 **Tartaglia, Niccolò,** 1499–1557.
Opere del famosissimo Nicolo Tartaglia cioè Quesiti, Nova Scientia, Travagliata Inventione, Ragionamenti sopra Archimede.
Venice, al segno di lione, 1606.

14.4 × 20.2 cm. Contemporary limp vellum.
viii + 284 pages + 48 + 52 + 72. General title-page, woodcut frontispiece, woodcut of instruments, 406 × 300 mm, 4 part title-pages with woodcut portrait of Tartaglia, *Quesiti* dedicated to Henry VIII, King of England, France and Ireland; *Ragionamenti* dedicated to Count Antonio Landriano (5 March 1551); *Nova Scientia* dedicated to Francesco Maria della Rovere, Duke of Urbino (20 December 1537); 58 woodcut figures in *Quesiti,* 33 × 35 to 92 × 96 mm; 9 woodcut figures in *Travagliata,* 45 × 16 to 88 × 170 mm; 12 woodcut figures in *Ragionamenti,* 14 × 43 to 108 × 75 mm; 34 woodcut figures in *Nova Scientia,* 48 × 77 to 141 × 202 mm.

Sixth Edition. Dedication to Luigi Giustiniano by Giovanni Battista Manassi, editor, is dated 22 March 1606.

REF: Manno, Cockle
COPIES: BM, MB, MiC, TU, OkU, IU, MCM.

THE FIRST EDITION OF *QUESITI* WAS PUBlished in Venice in 1554. It is in Book six of this work that Tartaglia writes about the form of urban fortification. The Newberry owns another copy of the *Quesiti* (14.5 × 20.6 cm, bound in reused parchment from a Hebrew manuscript of unknown date, 94 leaves-lacks signature A) whose publication date is unclear because the title-page is missing.

65 **Tensini, Francesco,** 1581–1638.
La fortificatione, guardia difesa et espugnatione delle fortezze esperimentata in diverse guerre.
Venice, Evangelista Deuchino, 1624.

25 × 35.5 cm. Damaged brown calf, goldstamped decorations and title in red morocco panel on spine. xvii + 83 pages + 83 + 128. Engraved title-page, engraved portrait of Tensini made in 1624 at 43, laudatory poems in Latin and Italian, table of contents, preface, subject index to Part One, 48 woodcut plates (32 full-page, 16 folded-in), 207 × 296 to 428 × 311 mm; plan of Vienna, engraved by Giovanni Giacomo Rossi in Rome 1683, 470 × 370 mm.

First Edition. Dedication to the Doge and Senate of Venice is dated 1 January 1624.

REF: Cockle, Miscellanea 14, Hale, de la Croix
COPIES: BM, BN (1630), DLC (1630), NWM, CU, DFo (1655), IU (1655), MiU (1655).

A NATIVE OF CREMA, TENSINI WAS INvolved in the military profession from the age of 17, first in Flanders in 1598 and 1601 in the service of the King of Spain, then in Frisia (1605), in Alsace and Bohemia—employed by Emperor Rudolf II—in Salzburg, Slessheim (with the Duke of Bavaria for five years), in Piedmont (1615) and in Friuli. His record was 18 sieges and 4 defenses, and his titles were engineer, captain, and lieutenant of artillery. He published this treatise while employed by the Republic of Venice, (from c. 1615) in whose service he fortified Crema, Bergamo, Peschiera and the citadel of Verona. He was knighted by the Republic, and the Senate approved in 1630 his plans for the fortification of Vicenza, not realized because of a popular rebellion.

This treatise, magnificently illustrated, is divided into three parts, of which the first one deals specifically with the design and construction of fortifications, as well as with the philosophical problems associated with the profession of military architecture. According to Tensini, military architecture is not pure philosophy but technology since it does not consist in demonstrable knowledge but is composed of principles needed for the defense and siege of cities. He contends that since ultimate truth is in power, military architecture is superior to its civic counterpart. He questions whether it is better to strengthen the frontiers or the center of the state, to fortify existing citadels or to build new citadels; he examines the quality of sites best for fortresses (listing plain, mountain, cliff, marsh, island, and peninsular locations) and discusses the principal parts of the fortress. His definition of the best fortress is one that will enclose the largest territory within the smallest circumference. Tensini insists on the use of earth rather than stone and brick in the construction of fortifications, even though he proposes the layering of the shooting platforms. Like de' Marchi he prefers wet moats, but criticizes the former's non-warlike use of them for swimming and fishing. Tensini refers to Alexander de Groote at end of book one in order to defend himself from accusations of plagiarism. It is rather de Groote who has learned from him—they were both in the service of the Duke of Bavaria—without, however, giving Tensini credit for his fortification method. The innovative elements of Tensini's fortification system consist of the *falsabraga*, the wet moat and the island-like ravelin.

66 Tetti, Carlo, c. 1529–1589.
Discorsi delle fortificationi.
Venice, Bolognino Zaltiero, 1575.

14.3 × 19.5 cm. Stained, yellowed limp vellum, two pairs of pigskin hook-and-button clasps.
viii + 119 pages + viii. Preface, 51 woodcut diagrams in text, 93 × 15 to 144 × 195 mm; 4 irregular woodcut plates folded-in, 276 × 194 to 392 × 305 mm, separate tables of contents and indices for each of the two books.

Second Edition, enlarged and corrected. Dedication to Maximilian II, Emperor of the Holy Roman Empire, dedication of part two to Rudolph, Archduke of Austria and King of Hungary.

REF: Cockle, Miscellanea 14
COPIES: BM (1569), MCM, BN, MB.

BORN INTO A PATRICIAN NEAPOLITAN family, Tetti studied with the Dukes of Zagarolo in whose house he lived. In 1550 he was on the African coast in Spanish service, and went to Vienna about 1565 where he consulted on the city's fortifications in 1575. He found Vienna ugly, badly located and lacking in resources. In 1576 he entered Venetian service, fortifying Verona and Bergamo.

The subject of the *Discorsi* had been discussed by the author with Pompeo Colonna, Duke of Zagarolo. The first edition of the *Discorsi,* published in Rome in 1569, came out unfinished and against the author's will, while the 1575 edition was enlarged to eight books and should be considered as a distinct work. Subsequent Italian editions were published in 1589 in Venice, 1585 in Rome, and 1617 in Vicenza.

The treatise is divided into 24 chapters. Among the themes treated are the siting of the fortress, the history of fortification, the elements of contemporary fortifications, and the new instrument for surveying invented by Tetti.

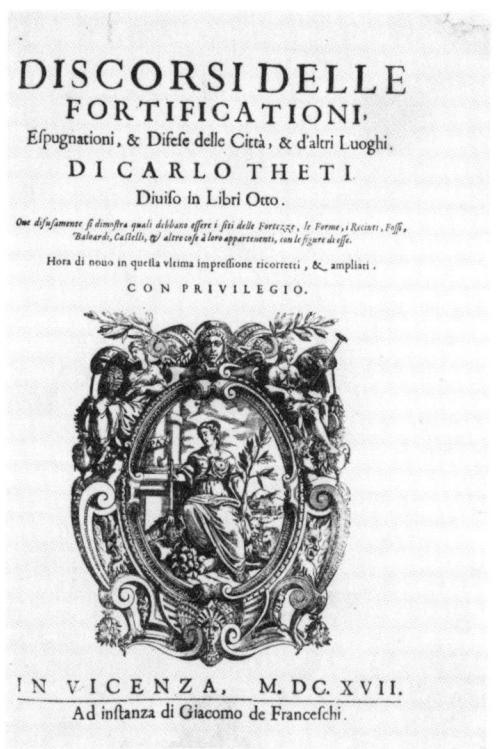

67 **Tetti, Carlo,** 1529–89.
Discorsi delle fortificationi, espugnationi, et difese della città . . . di Carlo Theti.
Vicenza, Giacomo de Franceschi, 1617.

25.2 × 36.2 cm. Repaired limp vellum.
iv + 210 pages. 248 woodcut diagrams, 91 × 12 to 504 × 362 mm.

Fourth Edition. Dedication to Nicolò Contarini, Venetian administrator of Friuli, is dated 10 March 1617 by the printer; dedication of part two to Rudolf, Archduke of Austria and King of Hungary by Carlo Tetti.

REF: Cockle, Miscellanea 14
COPIES: MH, MiU, BN, BM (1589), Bodleian (1589),

SIMILARLY TO THE 1575 EDITION, THIS later edition of Tetti's fortification treatise has been much enlarged. The chapters are abundantly illustrated with very coarse woodcut diagrams. According to Promis this is a reprint of a 1588 Venetian edition which he considers to have been post-mortem (thus really from 1589 as the BM and the Bodleian copies show).

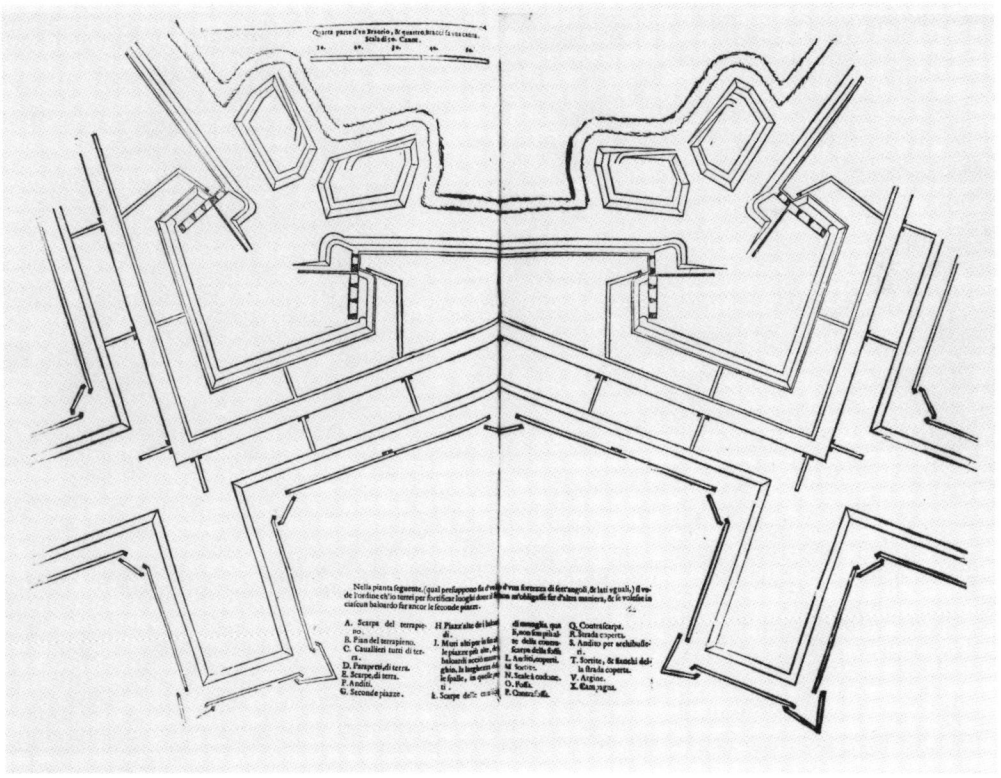

III

68   Valle, Battista della, d. 1535.

Vallo libro continente appertinentie à capitanij, retenere e fortificare una città con bastioni . . . e de espugnare una città . . . opera molto utile con le esperientia dell'arte militare.

Venice, Nicolo d'Aristotile detto Zoppino, 1529.

9.5 × 14.8 cm. Printed green and tan paper boards, marginalia in ink.

viii + 71 leaves + i. Woodcut title-page, table of contents, three chapters on fireworks, 20 woodcut illustrations of siege instruments and fortifications, 73 × 32 to 75 × 109 mm.

Fourth Edition. Dedication to Henrico Pandone, Count of Venafro.

REF: Cockle, de la Croix

COPIES: BM (1524, 1529, 1535, 1543, 1550, 1558), Bodleian (1535, 1558), NjP (1529, 1550), MiU (1529, 1550), NN (1529, 1539, 1543, 1558), IEN (1524), CSt (1524), PPF (1528), NcD (1528), CtY (1531), KU (1535), MnU (1550).

Battista della Valle seems to have been from Venafro and an experienced military man. His small "pocketbook" became one of the standard texts on the subject. Its popularity was demonstrated by the eleven editions printed in 37 years; it was more in demand than any other military book at the time. According to Cockle, della Valle had no scientific knowledge to speak of; he borrows heavily from ancient sources and relies upon his own experience in war.

The book is divided into four parts. Book one is what the captain needs to know about bastions and war machines, above all firearms, in order to defend the fortress; while book two discusses attack strategy and instruments. Book three is about the organization of the army for battle. Book four enters the fray of one of the most fundamental Renaissance debates—military arts versus letters—and illustrates it through conflicts, duels, fights, and backstabbing, finally promising the reader fame and honor:

> Se un'animoso core acquistar brama
> Nell'arte militar honor egreggio
> E far dell'opere volar la fama
> Nel quinto cielo al bellicoso seggio,
> Legga questo libretto, il qual si chiama
> Vallo, che vale ogni thesoro e preggio
> Per dar nell'arme (al buon soldato) lume
> Et adottarlo al martial costume.

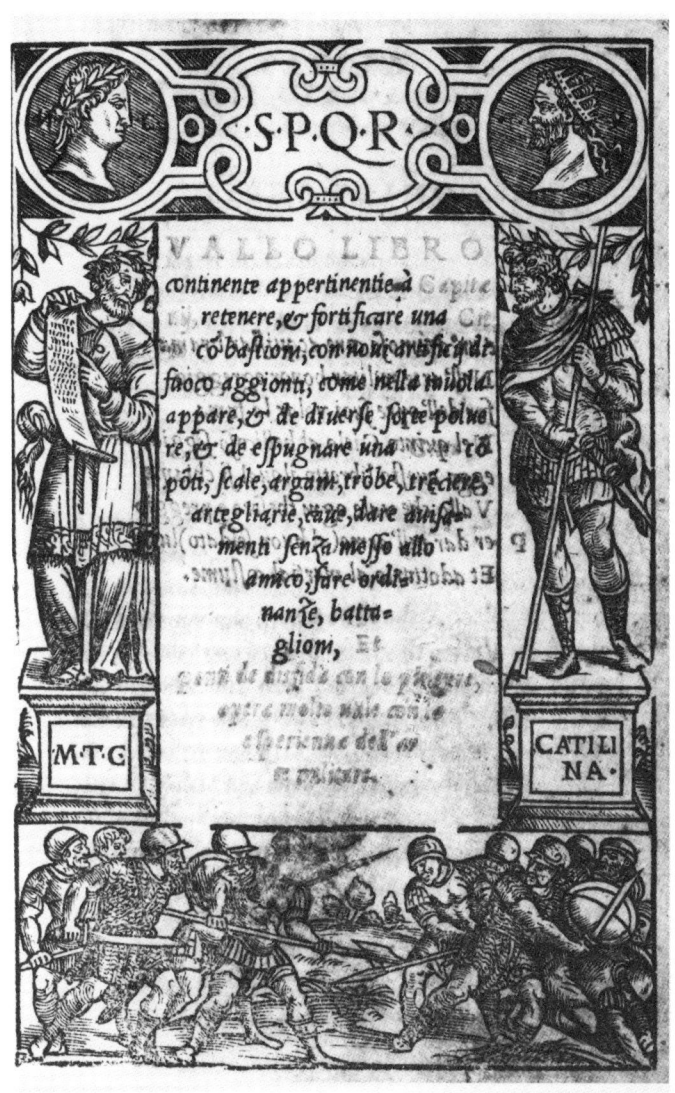

69 **Valle, Battista della,** d. 1535.
Vallo du faict de la guerre et art militaire.
Lyon, Jaques Moderne, 1554.

11 × 16.5 cm. Marbled paper boards, reinforced spine, gold edges.
iii + 87 leaves + i. Woodcut title page, table of contents, 20 woodcut illustrations, 68 × 28 to 90 × 143 mm.

First French Edition. Dedication to Henry Pandone Count of Venafro, license to publish dated 18 October 1554 in Lyons.

REF: Cockle, de la Croix.
COPIES: BM.

70 **Vauban, Sebastien Le Prestre de,** 1633–1707.
Le Directeur general des fortifications.
The Hague, Henry van Bulderen, 1685.

7.5 × 13.4 cm. Modern red cloth binding.
144 pages. Engraved title-page by A. de Blois, introduction.
Second Edition, pirated. Dedication to Prince of Orange.

REF: Architekt und Ingenieur, Blomfield, Blanchard, Grodecki, Parent and Verroust, Duffy
COPIES: HAB, BN, MH.

During his tenure as Intendant General of France's fortifications, Vauban supervised the reconstruction of about 160 fortresses. The construction ex-novo of 33 fortresses on the frontier of France with Germany, Flanders, and Spain ensured him lengthy travels. He served Louis XIV for 56 years and earned the rank of marshal. His interests ranged beyond fortification, attack and defense to city planning (through the *plans-reliefs*), strategy, agriculture, economy, and financial planning (*La Dîme royale*). He evolved the fortification methods inherited from French strategists and military architects such as Pagan, Deville and Clerville—his teacher—while also borrowing from earlier Italian sources such as de' Marchi. Vauban developed and realized three different fortification systems during his active career whose main idea, layering the defense in depth, was taken from Pagan. His most famous fortresses include Maubeuge, Lille, Neuf-Brisach, Longwy, Montlouis and Briançon.

The *Directeur General* and the treatise on attack strategy were the only publications composed by Vauban; the other treatises describing his work were written for him and after him. In the *Directeur* he describes the responsibilities of each member of the "corps de génie" in a succinct and comprehensive manner. Vauban was crucial in streamlining the operations of the French royal military engineers; the rules governing their recruitment, education and employment were established under his control. Among other topics, he discusses the functions of the director of fortifications, the intendant, the ordinary engineer, and the vanguard engineer. He provides formulas for the measurement of fortification materials and the estimate of expenses, and finally he explains the *devis* (the explanation of a fortification's detail, measurements and sequence of construction given to the builders, the equivalent of a modern blueprint), another of his controlling devices for the uniformization of fortification in France.

71 **Vauban, Sebastien Le Prestre de,** 1633–1707.
Science militaire contenant L' A. B. C. d'un soldat, L'art de la guerre et Le directeur general des fortifications.
The Hague, Adrian Moetjens, 1689.

7.1 × 13.4 cm. Brown calf, half title.
xxvi + 117 pages + xiv + 224 + vi + 144. Two engraved frontispieces, three part title pages, two tables of contents, 4 engraved figures in Part Two, 65 × 115 mm.

Part One, First Edition, dedicated to Count of Nassau, Governor of Bois-Le-Duc; Part Two, Fourth Edition, dedicated to Louis XIV, King of France; Part Three, Second Edition, dedicated to the Count of Orange.

72 **Vauban, Sebastien Le Prestre de,** 1633–1707.
The New Method of Fortification as Practiced by Monsieur de Vauban Engineer General of France.
London, Abel Swall, 1691.

11 × 18 cm. Worn calf, goldstamped title on red panel on spine, marginalia in ink.

xvi + 79 + 104 pages. Engraved title page, preface, table of contents, 8 geometrical diagrams and 22 fortification diagrams all double-page and folded-in, 210 × 180 mm. This copy has dishevelled and badly folded illustration plates and is rather worn.

First English Edition. Dedication to the Duke of Ormond.

REF: Architekt und Ingenieur, Blomfield, Parent and Verroust, Grodecki, Blanchard, Manno
COPIES: HAB (Amsterdam 1689, Paris, 1692), BN (Amsterdam 1734), DLC (1693), MH, MiU, OCU, NSyU (1691; French, Paris 1692), CtY (1691, 1693, 1702), DFo (1693, 1702), CLU-C (1691, 1693), MnU (1693), IU (1693), WU (French, Amsterdam 1689), NWM (French, Amsterdam 1689).

V<small>AUBAN'S WORK WAS PUBLISHED IN</small> both pirated and authorized versions during his lifetime. The title-page of this English edition is identical to the title-page of the 1694 Paris edition edited by Du Fay and de Cambray. Many of his realized projects could not be published for reasons of security, and for the same reason there was great interest in the fortresses designed by him.

73 Zanchi, Giovambattista de', 1515–c. 1586.

Del modo di fortificar le città.

Venice, Plinio Pietrasanta, 1554.

15.3 × 21.3 cm. Paper boards.
63 pages. Woodcut title-page, portrait of the artist, preface, introduction, 8 woodcut illustrations of fortification, 77 × 77 to 136 × 176 mm.

First Edition. Dedication to Maximilian of Austria, King of Bohemia.

REF: Cockle, Miscellanea 14
COPIES: BM (1554, 1560, 1601), HAB (Venice 1556), DLC, MnU, NN (1556), 1601), CtY (1556), OkU (1556), DFo (1556), MiU (1560, 1601), MH (1560), NNC (1560).

Giovambattista de' Zanchi was from Pesaro, and by 1543 held the rank of captain. In 1547 he was in Germany with the papal army led by Ottavio Farnese, but returned to Italy almost immediately. He took part in the Siena wars and then in those of the Carafà (1553–1557). In 1561 he went to Cyprus as engineer of the Republic of Venice for two years; then in 1565 he was sent to Ragusa. Zanchi was praised by Maggi, who had met him in Venice.

Zanchi was the first Italian to write a treatise exclusively on military architecture and the fortification of the city. He himself claims this distinction, and goes further to say that he is the first to write on this subject, evidently knowing nothing of foreign writers, such as Dürer. He expressly states that there are no other books on fortification; his explanation for this obvious lacuna is that veteran military men are reticent to share their hard-won knowledge with others who by reading their publications could easily plagiarize them. Even those who were willing to write about military architecture did not have enough experience because of the relatively recent introduction of the cannon.

In this brief work Zanchi summarizes a number of issues that were to become the principal themes of subsequent fortification treatises. Among them are a discussion of the means for taking a city (force, treason and treachery), an analysis of the strength of artillery, the form of traditional fortifications, the form of modern fortified cities and their constituent fortification elements, the comparison between fortresses in mountains and those in the plains, and the relative advantages of the attackers and the defenders of a fortress.

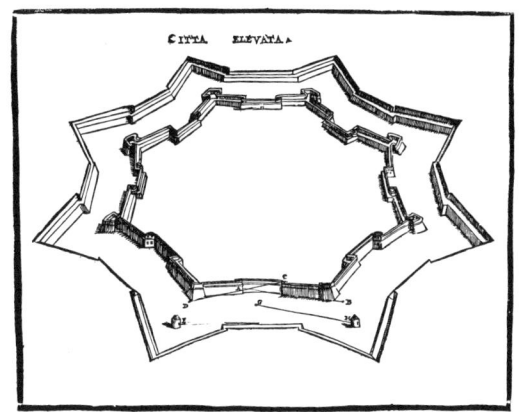
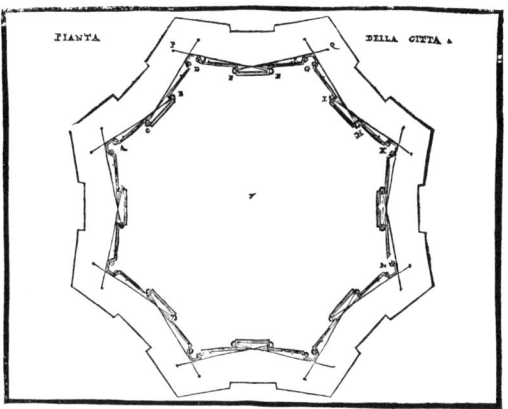